CAMBRIDGE
UNIVERSITY PRESS

Computer Science

for Cambridge IGCSE™ & O Level

PROGRAMMING BOOK FOR JAVA

Dave Duddell

CAMBRIDGE
UNIVERSITY PRESS

University Printing House, Cambridge CB2 8BS, United Kingdom

One Liberty Plaza, 20th Floor, New York, NY 10006, USA

477 Williamstown Road, Port Melbourne, VIC 3207, Australia

314–321, 3rd Floor, Plot 3, Splendor Forum, Jasola District Centre, New Delhi – 110025, India

103 Penang Road, #05–06/07, Visioncrest Commercial, Singapore 238467

Cambridge University Press is part of the University of Cambridge.

It furthers the University's mission by disseminating knowledge in the pursuit of education, learning and research at the highest international levels of excellence.

www.cambridge.org
Information on this title: www.cambridge.org/9781108910071

© Cambridge University Press 2021

First published 2021

20 19 18 17 16 15 14 13 12 11 10 9 8 7 6 5 4 3 2 1

Printed in Malaysia by Vivar Printing

A catalogue record for this publication is available from the British Library

ISBN 978-1-108-91007-1 Programming Book Paperback with Digital Access (2 Years)
ISBN 978-1-108-82419-4 Digital Programming Book (2 Years)

Additional resources for this publication at www.cambridge.org/go

Cambridge University Press has no responsibility for the persistence or accuracy of URLs for external or third-party internet websites referred to in this publication, and does not guarantee that any content on such websites is, or will remain, accurate or appropriate. Information regarding prices, travel timetables, and other factual information given in this work is correct at the time of first printing but Cambridge University Press does not guarantee the accuracy of such information thereafter.

Exam-style questions and sample answers have been written by the authors. In examinations, the way marks are awarded may be different. References to assessment and/or assessment preparation are the publisher's interpretation of the syllabus requirements and may not fully reflect the approach of Cambridge Assessment International Education.

The information in Chapter 12 is based on the Cambridge IGCSE, IGCSE (9–1) and O Level Computer Science syllabuses (0478/0984/2210) for examination from 2023. You should always refer to the appropriate syllabus document for the year of your examination to confirm the details and for more information. The syllabus document is available on the Cambridge International website at *www.cambridgeinternational.org.*

DEDICATED TEACHER AWARDS

Teachers play an important part in shaping futures. Our Dedicated Teacher Awards recognise the hard work that teachers put in every day.

Thank you to everyone who nominated this year; we have been inspired and moved by all of your stories. Well done to all of our nominees for your dedication to learning and for inspiring the next generation of thinkers, leaders and innovators.

Congratulations to our incredible winner and finalists!

WINNER

Patricia Abril
New Cambridge School, Colombia

Stanley Manaay
Salvacion National High School, Philippines

Tiffany Cavanagh
Trident College Solwezi, Zambia

Helen Comerford
Lumen Christi Catholic College, Australia

John Nicko Coyoca
University of San Jose-Recoletos, Philippines

Meera Rangarajan
RBK International Academy, India

For more information about our dedicated teachers and their stories, go to
dedicatedteacher.cambridge.org

CAMBRIDGE UNIVERSITY PRESS

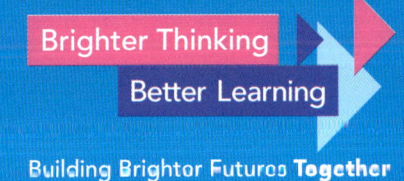

Brighter Thinking
Better Learning
Building Brighter Futures Together

> Contents

The items in orange are available in the digital edition that accompanies this book.

> Introduction

Study of the Cambridge IGCSE™, IGCSE (9–1) and O Level Computer Science syllabuses (0478/0984/2210) requires practical programming experience. Java is one of three programming languages recommended on these syllabuses, along with Python and Microsoft® Visual Basic.

The programming book for Java covers all of the specific programming requirements for the syllabuses. However, it could be used by anyone as an introduction to the Java language. The book is written with no assumption of any prior knowledge and there is no need for access to external reference sources.

Language

Java provides the normal functionality expected of a procedural or imperative high-level programming language. The syntax and constructs required for this functionality as detailed in the syllabuses are introduced in this book.

Java is also an object-oriented programming language. However, the syllabuses do not require study of object-orientation. This book therefore includes only minimal discussion of object-orientation. Coverage is limited to the object-oriented features needed to allow the running of a Java program containing the procedural language syntax and constructs.

Support

The book directly supports the syllabuses in the sense that chapters 1–11 cover subject matter which will help learners to work through the skills required for the Algorithms, Programming and Logic Paper.

To support learning and assessment, all the chapters include end-of-chapter tasks. Solutions to all the end-of-chapter tasks can be found on the digital part of this resource. There are examples of appropriate solutions that show how to turn logical ideas into actual programs. Finally, Chapter 13 provides thorough practice of exam-style questions. It will aid revision using knowledge from previous chapters.

Developing programming skills

A major feature of the syllabuses is the recognition that the creation of a successful program requires more than just the ability to write program code. There is emphasis on the need for the application of problem-solving and computational thinking skills. These can be used initially to identify the program requirements. Following on, suitable design techniques should be employed before program code is written.

The main chapters in this book support this approach. In particular, to support learning, these chapters include numerous tasks; both within the chapter and in the end-of-chapter questions. Most of these encourage the learner to apply computational thinking skills in addition to coding skills.

› How to use this series

The coursebook provides coverage of the full Cambridge IGCSE, IGCSE (9–1) and O Level Computer Science syllabuses (0478/0984/2210) for first examination from 2023. Each chapter explains facts and concepts and uses relevant real-world contexts to bring topics to life, including two case studies from Microsoft® Research. There is a skills focus feature containing worked examples and questions to develop learners' mathematical, computational thinking and programming skills, as well as a programming tasks feature to build learners' problem-solving skills. The programming tasks include 'getting started' skills development questions and 'challenge' tasks to ensure support is provided for every learner. Questions and exam-style questions in every chapter help learners to consolidate their understanding.

The digital teacher's resource contains detailed guidance for all topics of the syllabuses, including common misconceptions to elicit the areas where learners might need extra support, as well as an engaging bank of lesson ideas for each syllabus topic. Differentiation is emphasised with advice for identification of different learner needs and suggestions of appropriate interventions to support and stretch learners.

The digital teacher's resource also contains scaffolded worksheets for each chapter, as well as practice exam-style papers. Answers are freely accessible to teachers on the 'supporting resources' area of the Cambridge GO platform.

There are three programming books: one for each of the recommended languages in the syllabuses – Python, Microsoft Visual Basic and Java. Each of the books are made up of programming tasks that follow a scaffolded approach to skills development. This allows learners to gradually progress through 'demo', 'practice' and 'challenge' tasks to ensure that every learner is supported. There is also a chapter dedicated to programming scenario tasks to provide support for this area of the syllabuses. The digital part of each book contains a comprehensive solutions chapter, giving step-by-step answers to the tasks in the book.

> How to use this book

Throughout this book, you will notice lots of different features that will help your learning. These are explained below.

LEARNING INTENTIONS

These set the scene for each chapter, help with navigation through the Java subject material and indicate the important concepts in each topic.

SKILLS FOCUS

This feature supports your computational thinking, mathematical and programming skills. They include useful explanations, step-by-step examples and questions for you to try out yourselves.

QUICK QUESTION

This feature requests a brief pause for you to consider some aspect of the subject material under discussion. The question might request the identification of a fault or limitation in what has been presented and how it could be improved. This might require recall of some related knowledge you should already have.

KEY WORDS

Key vocabulary is highlighted in the text when it is first introduced. Definitions are then given in the margin, which explain the meanings of these words and phrases. You will also find definitions of these words in the glossary at the back of this book.

Pseudocode and Code snippets

This detailed method for describing the logic of computer programs will be very similar to the pseudocode used in the syllabuses.

```
INPUT day

WHILE day < 1 OR day > 31 DO

    OUTPUT "The value should be in the range 1 - 31
    please try again"

    INPUT day

ENDWHILE
```
Code snippet 10.1

TIPS

These are short suggestions to remind you about important learning points. For example, a tip to help clear up misunderstandings between pseudocode and Java.

Programming tasks

Programming tasks give you the opportunity to develop your programming and problem-solving skills. Answers to these questions can be found in the solutions chapter, on the digital part of this resource. There are three different types of programming tasks:

DEMO TASKS

You will be presented with a task and a step-by-step solution will be provided to help familiarise you with the techniques required.

PRACTICE TASKS

Questions provide opportunities for developing skills that you have learnt about in the demo tasks.

CHALLENGE TASKS

Challenge tasks will stretch and challenge you even further.

SUMMARY

There is a summary of key points at the end of each chapter.

EXAM-STYLE QUESTIONS

Questions at the end of each chapter which you can use to test your knowledge, understanding or skills. Some of these may require the use of knowledge from previous chapters. Answers to these questions can be found in the solutions chapter, on the digital part of this resource.

NOTE: As there are some differences in the way programming statements are structured between languages, you should always refer back to the syllabus pseudocode guide to see how algorithms will be presented in your exam.

⟩ Acknowledgements

The authors and publishers acknowledge the following sources of copyright material and are grateful for the permissions granted. While every effort has been made, it has not always been possible to identify the sources of all the material used, or to trace all copyright holders. If any omissions are brought to our notice, we will be happy to include the appropriate acknowledgements on reprinting.

Thanks to Getty Images for permission to reproduce images:

Cover Nerthuz; *Inside* Chapter 1 John Lund, Chapter 2 Andriy Onufriyenko, Chapter 3 Yuichiro Chino, Chapter 4 Andrew Brookes, Chapter 5 AniGraphics, Chapter 6 zf L, Chapter 7 schulz, Chapter 8 KTSDESIGN/SCIENCE PHOTO LIBRARY via Getty Images, Chapter 9 Nikada, Chapter 10 Rizky Panuntun, Chapter 11 Yuichiro Chino, Chapter 12 Nora Carol Photography, Chapter 13 Yuichiro Chino, Chapter 14 filo

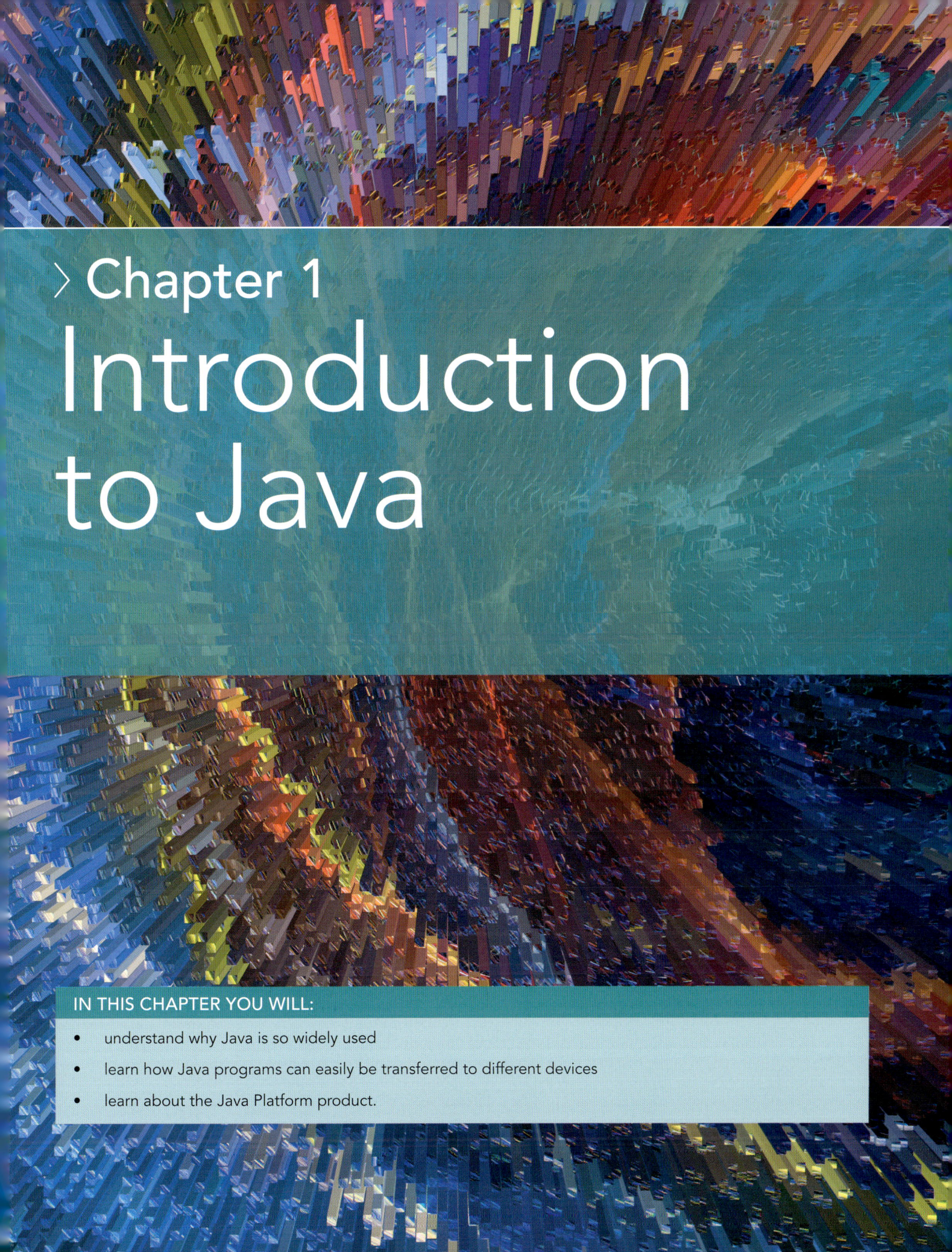

> Chapter 1

Introduction to Java

IN THIS CHAPTER YOU WILL:

- understand why Java is so widely used

- learn how Java programs can easily be transferred to different devices

- learn about the Java Platform product.

Introduction

There are three good reasons for choosing to learn to program in Java:

- For the beginner, Java is a straightforward language to learn.
- The software needed to write and run programs is free.
- Java is probably the most widely used programming language in the world.

At the time of writing, there are around 3 billion devices running Java, 12 million people developing Java programs and 25+ billion Java Cards in use in the world. (A Java Card is a smart card, like a credit card or a mobile phone SIM card, that has a microchip with software written in Java installed on it.)

This chapter explains the features that have made Java such an important programming language and includes details of how and why the language was created. The Java Platform is also introduced.

1.1 The origins and reasons for the development of the Java language

Java is based on the C programming language, which was developed in the 1970s. In the 1980s, the C language was expanded and transformed into a new language called C++. Java arrived in 1995. Java made use of many of the features of C++ but removed some of the more complex features. Sun Microsystems were responsible for the development and marketing of the early versions of Java. More recently, the product has been taken over by the Oracle Corporation.

The aim was for Java to be a language for writing programs that could be run on any hardware. At first, it was developed for use in things like fridges and washing machines; for example, a fridge has a **microchip** in it to monitor the temperature inside and switch the cooling unit on or off as required. However, people soon began to realise that the internet would provide more opportunities for the use of Java.

There are many different devices or systems that can be connected to the internet. It is therefore vital that it is easy to **port** software from one system or device to a different one. The porting of software involves transferring it to the new device and making the software usable there.

A program with good **portability** is one that can be transferred to many different types of device and installed on each device with minimum effort.

Java can be ported in the following way:

1 The program is written in Java.
2 A **compiler** is used to convert the program into **Java Byte Code**.
3 The Java Byte Code is transferred to the target hardware.
4 A **Java Virtual Machine** already installed on the hardware translates the Java Byte Code, allowing the program to run.

The Java Virtual Machine is software that has to be created for specific hardware. However, there is only one version of the Java Byte Code for any particular program. This Java Byte Code can be ported to any hardware that has the Java Virtual Machine installed. As the alternative to this portability is to create and transfer a different specific executable file for every different type of hardware, you can see that Java's portability is preferable.

KEY WORDS

microchip: a solid-state device with installed software.

port: a word used to describe the transfer and installation of software from one system or device to a different system or device.

portability: a measure of how easy it is to transfer software from one system or device to a different system or device and make it usable.

compiler: software used to translate program code into an intermediate code.

Java Byte code: an intermediate code for a program that can be ported to different hardware devices.

Java Virtual Machine: software installed on a hardware device that runs a program by using the Java Byte code.

1.2 The Java Platform

In learning to program, you will need to use only a small part of the programming language that experienced developers use. However, you will find it useful to know some of the terms used to talk about Java.

The **Java Platform** is the product that includes the Java language together with a variety of software to be used to support program development.

There are three different versions of the Java Platform:

- One version is for software to be used on a server.

- One version is for mobile devices.

- The third version is **Java Standard Edition (JSE)**. This is the one you will use because it is suitable for developing a program on a PC.

The JSE includes the **Java Development Kit (JDK)**, which contains the compiler. With this you can write programs and convert them to Java Byte Code. The JSE also contains the **Java Runtime Environment (JRE)**, which, in turn, contains the Java Virtual Machine. This allows a program to be run.

JRE is often installed by itself when the only requirement is to be able to run a Java program.

Figure 1.1 shows the relationship between the most important parts of the Java Platform.

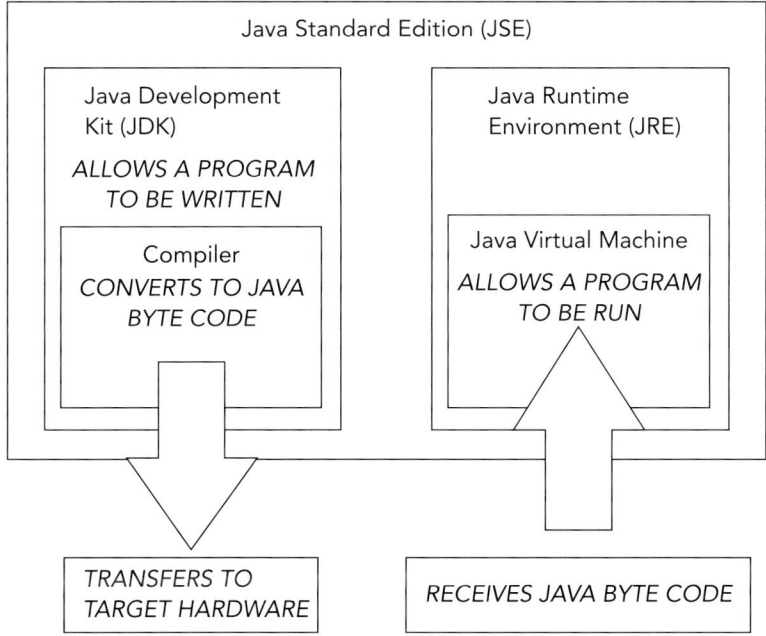

Figure 1.1: An overview of how the parts of Java Standard Edition are used

The product is often updated. Fortunately, the changes very rarely affect the language features that you will be using as you begin to learn how to program. At the time of writing, the latest version is Java SE 13. Some earlier versions are also still supported.

QUICK QUESTION

Do you know which version of Java you expect to use? If you are unsure, you could visit the web pages provided by Oracle.

SUMMARY

Java is the most widely used programming language in the world. Its success is due to the portability of Java programs.
Java provides software to run on devices that are connected to the internet.
Portability allows you to easily transfer programs to other types of device. It is achieved through the use of Java Byte Code and the Java Virtual Machine.
The product that is provided is called a Java Platform.
Java Standard Edition (JSE) is the version of the Java Platform that is used for developing programs to be run on a PC.
The two major parts of the Java Platform are the Java Development Kit (JDK) and the Java Runtime Environment (JRE).

END-OF-CHAPTER QUESTIONS

1 State what code is used by the Java Virtual Machine.
2 Describe the different uses of JDK and JRE.

> Chapter 2

Algorithms

IN THIS CHAPTER YOU WILL:

- understand the concept of an algorithm
- learn how the three logic constructs (sequence, selection and iteration) can be used in an algorithm
- understand that a preliminary algorithm design can be presented in Structured English
- learn that a detailed algorithm design can be created using pseudocode or a flowchart
- learn about the components used to construct a flowchart
- create a preliminary design using a flowchart.

Introduction

The aim of this book is to support you while you are learning how to write computer programs using Java. This chapter introduces some concepts that you need to consider before you begin to write program code. These concepts apply whichever programming language you use.

First of all, you need to understand that a program can be described as an algorithm – a series of actions to achieve a specific outcome. You then need to be familiar with methods of presenting an algorithm design. The following statements sum up the reasons why the content of this chapter is important:

- Actually writing a computer program should be seen as the last step in a problem-solving process.

- Unless the problem is very simple, a solution should be documented as an algorithm design before it is implemented as a program. A documented design has all of the details recorded so that there can be no misunderstandings when coding begins.

- You need to be familiar with methods that are used to document a design.

- In the chapters dealing with different elements of the Java language, you will be provided with designs for problem solutions. These may be presented together with the corresponding Java code or with a request for you to write the code.

2.1 Algorithms

There are different ways that a computer can be used to solve a problem. In this book, you are going to learn about a particular approach. In this approach, a program is designed, then implemented by writing program code using a high-level programming language. This is a type of language that tells the computer what it must do and how it should do it. Java is one of many languages of this type.

A program written in Java will consist of one or more **algorithms**. An algorithm can be defined as a series of actions. These actions must follow a logic that is suitable for solving a problem.

We do not use the term 'algorithm' much in daily life. However, there are many examples of activities in our daily routine that follow a series of actions – in other words, an algorithm. Examples include making a cup of tea, following a cooking recipe, repairing a tyre. The following is a suggestion for an algorithm to be used when two players are going to play chess.

> **KEY WORD**
>
> **algorithm:** a series of actions required to achieve a specific outcome.

1 Place the chessboard on a table with a white square bottom right.
2 Place the pieces on the board.
3 Choose who is to play white.
4 White makes the first move.
5 Players take turns to make a move.
6 Play continues until there is a final result.

There are several points to note here:

1 It is not essential to number the actions. However, using the numbers highlights the fact that, in this example, the actions must be carried out one after another.

2   There is a lot of detail missing, such as how the chess pieces are arranged on the board and how a final result is achieved. We will discuss this later in the chapter.

3   The order in which the actions in the algorithm are performed must follow a logic that is suitable for solving the problem.

4   The version shown here does have a slight problem with the logic.

2.2 Structured English

In Section 2.1, a full English sentence was used to define each action in the chess algorithm. This approach could be used as a first step in producing a design for an algorithm for a computer program. However, it would take a long time to write it, and then a long time to read it when the next stage in the design process began.

An alternative is to use **Structured English**. Here is an example for the chess algorithm discussed in Section 2.1:

> Place board on table.
>
> Place pieces.
>
> Choose who is white.
>
> White moves.
>
> Players take turns.
>
> Continue until final result.

Each action is now described using English words, but using as few as possible. There is no defined way of writing Structured English. In Section 2.3, Structured English is used to define the design for some algorithms that illustrate the logic constructs that can be used in an algorithm. When you come to design a program for yourself, you may find it useful to begin with a Structured English representation of the design.

PRACTICE TASK 2.1

Consider the order of the actions in the chess-playing algorithm design in Section 2.2. Identify the problem with the logic. Suggest an improved version by rearranging the order of the actions.

2.3 Program logic constructs

A **logic construct** controls the order in which algorithm actions take place. There are three options: sequence, selection and iteration.

Sequence

It is possible for an algorithm to consist of a simple **sequence** of actions where each action happens once and there are no alternative routes available. The chess problem in Section 2.2 is an example.

As another example, we can consider the design of an algorithm for a program that a teacher might use. This simple program takes a score achieved by a candidate in an exam and converts it to a percentage mark.

Algorithm 1

The design could be written in Structured English as follows:

> Input total mark possible.
>
> Input candidate mark.
>
> Calculate percentage.
>
> Output percentage.

DEMO TASK 2.1

You have been asked to write a program that will calculate the number of tins of paint that would be needed to paint a wall. You are not ready to start programming yet, but you are going to consider a design.

Solution

The first thing you need to do is to think about the problem and decide what data you would need for the calculation. The following would be a sensible list:

- The length of the wall.

- The height of the wall.

- The area that one tin of paint would cover.

- How many coats of paint you would expect to use.

You can see that you will need to calculate the area of the wall before you can calculate the number of tins needed. The following is, therefore, a sensible Structured English design for your program:

> Input length.
>
> Input height.
>
> Calculate area.
>
> Input area covered by one tin.
>
> Input number of coats.
>
> Calculate number of tins needed.
>
> Output number of tins needed.

QUICK QUESTION

Would there be any problem if there was a different order of the tasks in the above design?

Selection

As an example to illustrate this construct, we can return to considering a program for a teacher. Suppose the teacher now wants a program that converts the percentage mark to a grade. It is no longer possible for the algorithm to consist of one simple sequence.

Algorithm 2

The following could be a suitable design:

Input total mark possible.

Input candidate mark.

Calculate percentage.

Grade as Distinction if percentage above 80.
Grade as Merit if percentage in range 61–80.
Grade as Pass if percentage in range 40–60.
Grade as Fail if percentage less than 40.
} Selection items

Output grade.

This algorithm illustrates the use of the **selection** construct. You should note that there is a **condition** being tested to control which action happens (the choice of grade depends on the candidate percentage). In this example, the use of the selection construct does not affect the logical flow of the whole program. In other examples, there might be two alternative sections of the algorithm that could be followed. The result from testing the condition chooses which one is followed.

Iteration

Algorithm 3

So far, the program is only providing a grade for one candidate. If the teacher needs a program to output a grade for every candidate, then a simple change can be made to the design, as follows:

Do the following for each candidate: } Iteration

Input total mark possible.

Input candidate mark.

Calculate percentage.

Grade as Distinction if percentage above 80.
Grade as Merit if percentage in range 61–80.
Grade as Pass if percentage in range 40–60.
Grade as Fail if percentage less than 40.

Output grade.

This illustrates the use of **iteration**, which is the last of the three logic constructs. This is where a group of actions is repeated several times.

There are two points that are worth noting at this stage. The first is that there is a lack of detail in these designs. For example, there is no formula included for the calculation of the percentage. The second is that indentation has been used to help the reader to follow the logic. Indentation is where the start of the text on a line of the design is moved away from the left-hand margin. The example in Algorithm 3 shows the use of two levels of indentation.

KEY WORDS

selection: a logic construct in which alternative algorithm actions are possible with the choice dependent on testing a condition.

condition: the criteria that are tested which provide either TRUE or FALSE as an answer.

KEY WORD

iteration: a logic construct in which a sequence of actions in an algorithm is repeated several times.

TIP

Iteration is a rather formal word. Many people will refer to this as 'looping'.

PRACTICE TASKS 2.2–2.3

2.2 For all of these designs, it is possible that the teacher would find the program more useful if something else were output as well as the percentage or grade. Identify a sensible additional output, then rewrite one of the designs to include this extra output. You might need an extra input as well!

2.3 The Structured English in Algorithm 3 does not have the most sensible logic. Rearrange the statements to improve the logic.

CHALLENGE TASKS 2.1–2.2

2.1 A program is needed to work as a calculator. The user is asked to input two separate numbers and an arithmetic operation (+ , × or −). The program must then calculate the answer. This answer must be output.

Create a Structured English design for this program including suitable input and output.

2.2 Consider again the design in Algorithm 2. Identify one or more reasons why the teacher might want to add to the algorithm to extend what the algorithm does. Explain your reasons. Provide a Structured English design that includes the extra features.

2.4 Design tools

As we have seen in the previous examples, a programmer will sometimes be writing a program requiring only a few lines of code. At the other extreme, a large organisation may, from time to time, require the development of a much more complex software-based system.

For the complex software-based system, the development team will have a wide variety of design tools available, from which they can choose the most suitable. By contrast, a program with few lines of code can be written without any use of design tools.

If there is going to be a design process, then in the final stage a **detailed design** must be documented. The amount of detail in this detailed design should be sufficient for the programming to begin.

In Section 2.3, three designs are presented in Structured English for an algorithm to be used by a teacher. These designs illustrate the program logic constructs that are used in an algorithm (sequence, selection and iteration). You are not expected to use Structured English for a detailed design. However, it is useful to consider the level of detail included in these designs and whether there is sufficient detail for programming to begin. There are three aspects to consider:

1   Each algorithm requires the calculation of a percentage. Such a simple calculation probably does not need to be further detailed in the designs.

2   Algorithms 2 and 3 require a grade to be given depending on the percentage mark. The detail provided for this is essential because each teacher will have their own specific set of rules for deciding on a grade.

TIP

In the context of software development, you will often come across a reference to a system or to a program. It is best to consider these as having the same meaning.

KEY WORD

detailed design: the final product of a design process that contains enough detail for programming to begin.

3   Algorithm 3 involves an iteration. A detailed design would need to specify how this should be implemented in the algorithm.

Figure 2.1 shows the stages of design leading to program coding. It shows that there are two techniques available for detailed design: pseudocode and flowcharts.

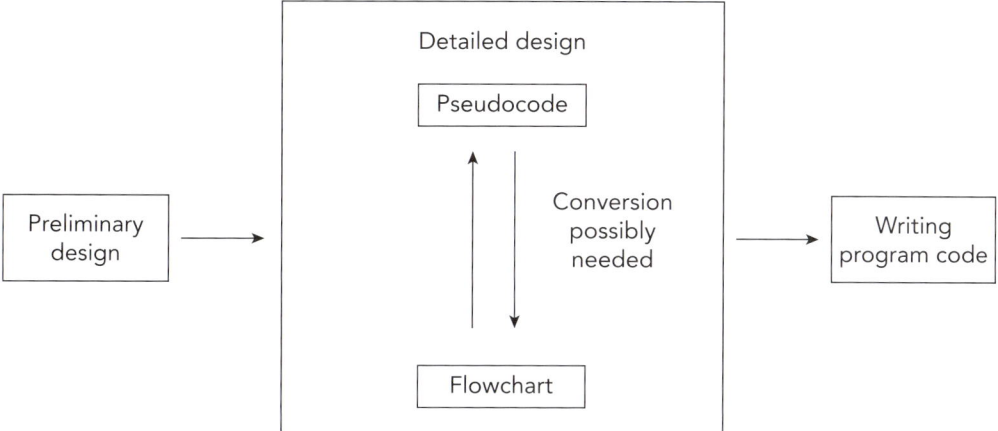

Figure 2.1: Stages of design leading to program coding

This section provides some comments about pseudocode and some details regarding flowcharts.

Pseudocode

Pseudocode is a way of representing the sequence and logic of a program using natural language words, a bit like Structured English but much more detailed. You can use pseudocode to provide a design that has all of the coding components that would be used in an actual programming language. Therefore, a design presented in pseudocode looks like a program code listing. The design can be understood by programmers of any language.

This book will introduce examples of pseudocode in each of the following chapters alongside the corresponding Java code.

KEY WORD

pseudocode: a way of unambiguously representing the sequence and logic of a program using both natural language and code-like statements.

Flowcharts

A **flowchart** is a graphical representation of a detailed design for an algorithm. It uses a set of symbols, which are illustrated and defined in Table 2.1.

KEY WORD

flowchart: a diagram used to document a detailed design for an algorithm that shows the logical flow of the actions.

Symbol	Use
(rounded rectangle)	The terminator symbol: there is one at the start and one at the end (referred to as 'STOP') of the flowchart sequence.
(parallelogram)	An action symbol: it is used to show that the action required is either an input of a value or the output of a value. There must be only one arrow going in and one arrow going out.
(rectangle)	An action symbol: sometimes referred to as the 'process symbol'. It is where one or more actions can be defined. There must be only one arrow going in and one arrow going out.
(diamond)	An action symbol: referred to as the 'decision symbol'. It defines a condition to be tested. There must be only one arrow going in but there must be two arrows coming out. One of the outgoing arrows must be marked TRUE or T and the other marked FALSE or F.
(arrow)	The arrow is used to define the sequence or flowline of the actions defined in the flowchart.

Table 2.1: Flowchart symbols

There are a couple of points to make about the use of flowcharts:

1 A flowchart is normally presented as a design going from top to bottom, rather than from left to right, with the start terminator at the top and the stop terminator at the bottom.

2 A flowchart is normally used for a detailed design. In this case, each of the three action symbols must contain specific details of the action. Pseudocode is suitable for this.

In the following chapters, when a detailed design is provided for you, this will usually be presented as a flowchart and as pseudocode.

Flowchart 2.1 shows an example of a flowchart constructed as a preliminary design for a program that provides a quiz for a user. Note how iteration is included by the use of an arrow returning to an earlier stage in the algorithm.

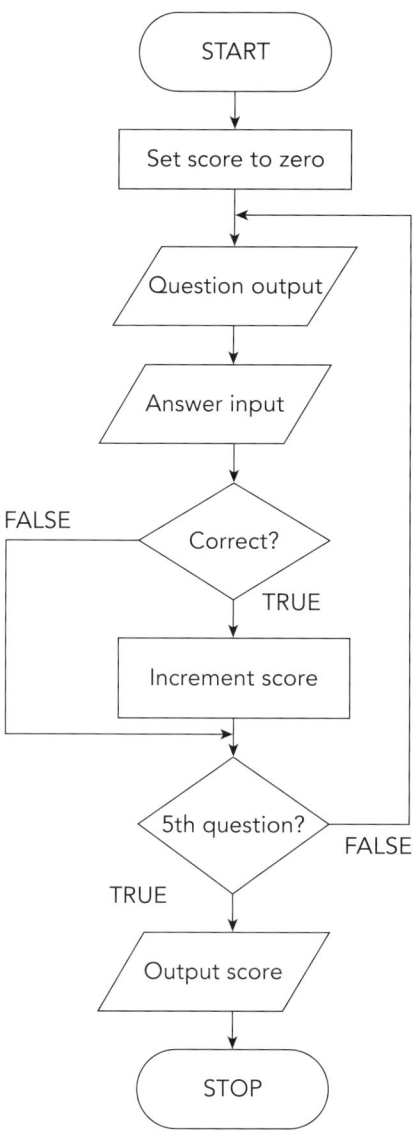

Flowchart 2.1: A flowchart for a program that provides a quiz for a user

The overall logic is defined here in this preliminary design. Note the need to **initialise** the value for the running score to zero before any questions are asked. This score is increased only if the user provides a correct answer. Only five questions are asked.

The design does not consider how the questions and the answers are stored in the program.

KEY WORD

initialise: an action that provides a value for a variable before the variable is first used in a program.

DEMO TASK 2.2

*You wish to create a program that adds up a series of values that are input. You decide to create a flowchart as a **preliminary** design. At this stage, you are going to use Structured English to label the flowchart symbols. (In later chapters you will be using pseudocode.)*

CONTINUED

Solution

After considering the requirement, you realise that a loop (iteration) will be needed. In each iteration, the newly input number must be added to a running total. For this to work properly, the total must be set to zero before the loop begins.

Flowchart 2.2 shows the flowchart for the preliminary design.

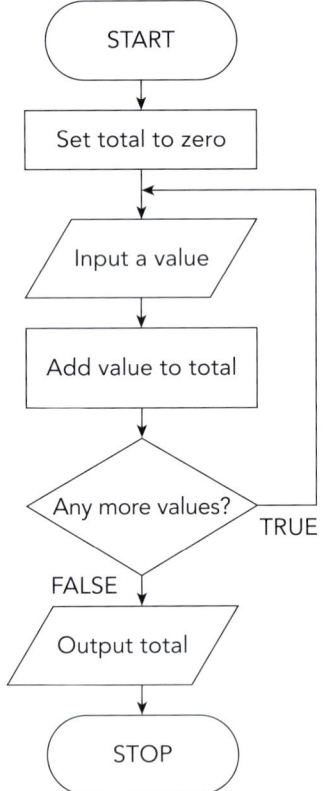

Flowchart 2.2: A preliminary design flowchart for an algorithm to add up values

PRACTICE TASK 2.4

Create a flowchart for a program that could be used by a teacher. You can use the Structured English design for Algorithm 1 provided in Section 2.3. You are not asked to provide pseudocode to define the actions.

CHALLENGE TASK 2.3

In Challenge Task 2.2, you created a Structured English design for a program to be used by a teacher. The design had some added features to make it more useful. You now need to document that design as a flowchart. You can use Structured English from your design to label the symbols.

SUMMARY

An algorithm consists of a series of actions.
An algorithm uses logic constructs to define the order in which the algorithm actions are performed.
The three logic constructs are: sequence, selection and iteration.
Structured English can be used to define an initial design for an algorithm.
Structured English contains brief statements, not full sentences.
The final stage of the design process should be the creation of a detailed design.
There should be sufficient detail in a detailed design to allow program code to be written.
A detailed design for an algorithm can be presented as pseudocode or as a flowchart.
Pseudocode uses code components similar to ones used in a programming language and can be read and understood by programmers of any programming language.
A flowchart is a diagram which shows the order in which actions are performed in the algorithm.

END-OF-CHAPTER QUESTIONS

1   Give definitions for the following terms:

   i   an algorithm   ii   a logic construct   iii   sequence

   iv   selection   v   iteration.

2   State what the following flowchart symbols are used for:

3   Consider the following scenario:

   A shopkeeper needs a program to calculate the total bill for items bought by a customer.
   The program is to work as follows:

   •   The following will be repeated for each different item bought:

      1   The price for one item will be input.

      2   The number of that type of item bought will then be input.

      3   The price will be calculated for the number bought of this item.

      4   This price will be added to a running total.

   •   The total price to pay will be output.

   For example, if you were to buy five pencils and some other items:

      1   The price of one pencil will be input.

      2   The input for the number of pencils bought will be 5.

      3   The price for these five pencils will be calculated.

      4   That price is added to a running total.

   Construct a flowchart for this program. You can use Structured English to label the symbols.

> Chapter 3
Variables and arithmetic operators

IN THIS CHAPTER YOU WILL:

- learn about the use of variables in a high-level programming language
- define data types for variables
- understand the difference between a variable, a constant and a literal
- learn how arithmetic operators are used in assignment statements
- learn about the use of an identifier table.

Introduction

A variable is an identifier that is used when a value is stored in memory. This chapter explains why variables are used in a high-level language program. There is a discussion about the coding that is used for the creation and use of variables in a program including the pseudocode that would be used in a design. The focus is on arithmetic using variables. Finally, there is an introduction to the use of an identifier table as a design tool.

3.1 Using variables in a high-level programming language

Suppose that you create a simple program to find out how much it would cost to purchase an item online and have it sent to your house. You would need to input two values, calculate the sum of the two values, store and then output this sum. The following would be the sequence of actions that would take place when the program was run:

1   The first value is input and stored in memory (the price of the item).
2   The second value is input and stored in memory (the price of posting and packaging).
3   The two values are retrieved from memory and added together.
4   The value calculated for the sum is stored in memory.
5   The value calculated for the sum is output (the total price you pay).

If the program was written in a low-level language, the program would have to identify a memory location every time a value was stored in memory. It would then have to remember that memory location every time that a value was retrieved from memory.

By contrast, if the program is written in a high-level language, such as Java, the programmer can create a **variable** to identify a value to be used in the program. The programmer does not have to worry about the memory locations used. A translation program will handle these. The programmer writes the **source code**. Then the translation program will use the source code to create a corresponding low-level program that the computer can **execute**.

3.2 Variable types

If a low-level program makes a mistake in identifying a memory location, the program will continue to run using the incorrect data value. This is because all memory locations just contain binary digits. There is nothing stored in memory to indicate what these binary digits represent.

When a programmer uses a high-level language to write source code, the language forces the programmer to choose a **data type** for every variable. For each data type, the following are defined:

- The range of values that are allowed for the variable.
- The number of memory locations used to store a single value.
- The different ways that a variable of the particular type can be used.

KEY WORDS

variable: a named memory location used to store a value. The value of the data can be changed during program execution.

source code: a program written by a programmer using a high-level programming language.

execute: in Computer Science, the term 'execute' means the operation of a computer program. When a computer program is in operation it is being executed. The term 'run' is also used to describe the same process. The 'program is running' or the 'program is being executed' both mean the same thing.

data type: a specification of the kind of value that a variable will store.

The use of a data type gives the opportunity to find errors in a program before any attempt is made to run the program.

Table 3.1 shows the five data types you need to be familiar with.

Generic name	Java version	Pseudocode version
Integer	int	INTEGER
Real	float	REAL
Character	char	CHAR
String	String	STRING
Boolean	boolean	BOOLEAN

Table 3.1: Data types for variables

The *Generic name* column identifies data types that you would expect to be available for any high-level programming language. The *Java version* column identifies the specific words that are defined in the **syntax** of the Java programming language. These are the words you need to use when writing Java. You should note that Java is a case-sensitive language. You must always write the type exactly as written in the *Java version* column.

The *Pseudocode version* column identifies the format that will be used when you are presented with pseudocode. This is explained in Section 3.3.

Table 3.2 contains some information about the Java types that are introduced in Table 3.1.

Java type	Relevant information
int	Whole number values only, positive or negative in the range −2147483648 to 2147483647
float	Positive or negative numeric values with a fractional part
char	A single Unicode character
String	A chosen number of Unicode characters
boolean	The value is either true or false

Table 3.2: Explanation of data types

3.3 Declaring variables

Java follows the convention that applies to nearly all high-level programming languages. It is necessary for a variable to be included in a **declaration statement** before it can be used in a program. In a declaration statement, the name of each variable is created and its data type defined.

The following are some examples of declaration statements:

```
int firstNumber;
float averageForPaper1, totalAverage;
boolean candidatePresent;
char grade;
```

You can see that each statement begins with the data type followed by a space. This is followed by either the name of a variable or a list of names for several variables separated by commas. The statement finishes with a semi-colon.

You need to make a sensible choice for the name of each variable. The following are some points to remember.

1   You must not choose a Java **keyword** as a variable name. (A list of keywords is given in the Appendix.)

2   Java does not allow the first character of a name to be a numeric character.

3   It is best to use an alphabetic character as the first character.

4   For the remaining characters, there are in effect no restrictions.

5   The name chosen should relate to how the variable is used.

6   **camelCase** is recommended where the name consists of two or more words. camelCase is where a capital letter is used for each word except for the first word. An example is `firstNumber`. This is one of several used in the example declaration statements provided earlier in this section.

7   Whatever choice you make, you must remember that Java is case-sensitive. Every time you use a variable, the name must be written each time with exactly the same choice of upper or lower case for each alphabetic character in the name.

Pseudocode

Chapter 2 contained a brief discussion about pseudocode and its use for providing a detailed design for an algorithm.

There is no universally agreed standard syntax for pseudocode. However, Cambridge International have defined a syntax in the syllabus which will be used in exam questions. Variable names will use mixed case.

In this book pseudocode will use mixed case in the camelCase style described above for Java code which is with the first character in lower case.

This is illustrated by the examples used in the following declaration statements:

```
DECLARE firstNumber : INTEGER
DECLARE averageForPaper1, totalAverage : REAL
DECLARE candidatePresent : BOOLEAN
DECLARE grade: CHAR
```

The most important features to note here are the use of upper case for keywords and the lack of a semi-colon at the end of a statement.

TIP

Documentation about programming languages often uses the word 'identifier' rather than 'name'. 'Identifier' and 'name' have the same meaning in the documentation.

KEY WORDS

keyword: a word with a specific meaning defined by a programming language.

camelCase: a way of creating a variable name from a combination of at least two words. Each new word after the first word starts with a capital letter.

DEMO TASK 3.1

As you can see, programmers have two decisions to make when choosing variables for a program. The first is the choice of sensible data types. The second is the choice of sensible names. This can be illustrated by considering the following problem.

> CONTINUED
>
> *Children are going on a school trip. They have been asked to select what they want for lunch. The teacher needs a program to calculate the total cost of the meals. Each child has to choose one from each of the following options:*
>
> - *Sandwich or burger.*
>
> - *Fruit or cheese.*
>
> - *Water or juice.*
>
> Solution
>
> Once you have thought about the problem, you can see that you will definitely need:
>
> - Variables to store real values for individual prices.
>
> - Variables to store integer values for the numbers chosen for each item.
>
> - A variable to store the total cost as a real value.
>
> On the basis of this, you decide on the following declaration statements that you will use in your program.
>
> ```
> float sandwichPrice, burgerPrice, fruitPrice;
> float cheesePrice, waterPrice, juicePrice;
> int sandwichNumber, burgerNumber, fruitNumber;
> int cheeseNumber, waterNumber, juiceNumber;
> float totalPrice;
> ```

> PRACTICE TASK 3.1
>
> Consider a program that is to be used to record information about members of a group. It is required that names and ages will be used when the program runs. What variables can you use for the name and the age?
>
> Try to think of at least two variables each to store values for the name and for the age. When you have made a decision, write declaration statements in Java for the variables you have chosen. You should aim to make the variable names as meaningful as possible.

3.4 Variables, constants and literals

There are three ways that a value can be included in a program.

The first way is to associate the value with a variable.

The second way is to use a **constant**. You use a constant when a particular value is to be used several times in the same program. A constant has to be declared. For Java, an example of a declaration statement for a constant is:

```
final int maximumPercentage = 100;
```

Table 3.3 contains an explanation for each component of this statement.

Component	Explanation
`final`	This is a keyword that makes the statement a declaration for a constant.
`int`	This defines the type for the constant.
`maximumPercentage`	This creates the name for the constant (uses camelCase).
`=`	This supplies a value for the constant.
`100`	This is the value.
`;`	This is essential to finish the statement.

Table 3.3: Explanation of a declaration statement for a constant

This constant declaration would be presented in pseudocode as:

```
CONSTANT maximumPercentage ← 100
```

Note that the value is given as soon as the constant is introduced. You are then not allowed to change the value later in the program.

The third way to include a value is to use a **literal**. A literal is a value that you use directly in the program without any previous declaration. The 100 explained in Table 3.3 is an example of a literal. A literal value does not require a formal identification of its type. You simply write the literal in a way that identifies its type. For example, 12 would be used as an `int` value but 12.0 would be used as a `float` value.

A `String` literal is identified by the use of double quotes. For example, "Hello" could be used in a welcome message to a user of a program. It is important to understand that the quotes are only there to define the start and finish of the string. The quotes are never displayed.

A `char` literal uses single quotes around the single character. An example is 'a'.

3.5 Initialising variables

When you first supply a value for a variable, you are said to **initialise** the variable. It is good practice to do this early in the program. If a program uses a variable that has not been initialised, the results may not be predictable.

There are three ways that you can initialise a variable in a program:

1   The program can **assign** a value in a declaration statement as was shown for a constant declaration in Section 3.4.

2   The program can request a value to be input by a user of the program. Chapter 4 will show how this is done in Java.

3   The program can use an assignment statement, as discussed in Section 3.6.

KEY WORD

literal: a value used directly in a program.

TIP

You should always choose to define a constant if a value is going to be used more than once in a program and will never be changed.

KEY WORDS

initialise: an action that provides a value for a variable before the variable is first used in a program.

assign: a word that applies specifically to supplying a value to a variable.

3.6 Assignment statements and arithmetic operators

The **assignment statement** is the most important concept in high-level programming. It can be described as:

A variable is given a value obtained by evaluation of an **expression**.

For Java the format is:

```
<Variable name>  =  <expression>;
```

The left-hand side of the statement must consist of just one variable. The = sign does not represent equality. Instead, it represents the action of giving a value to the left-hand side variable. The right-hand side expression must contain at least one value but can contain more. If the expression has more than one value, at least one operator will be included to define how those different values are used.

In this chapter, we are only going to consider the use of an **arithmetic expression** in an assignment statement. We will look at other forms in later chapters. In this case, the assignment statement assigns a numeric value to a variable.

The simplest form of such an assignment statement is where the expression is a literal. The following is an example in Java:

```
maximumNumberOfEntrants = 100;
```

The same assignment if presented in pseudocode would be:

```
maximumNumberOfEntrants ← 100
```

The use of the back-arrow symbol in pseudocode for the assignment action emphasises that there is no concept of equality here.

Many assignment statements involve the evaluation of arithmetic expressions containing more than one numeric value. A value in the expression might be provided as a literal. The alternative is to use a variable that has previously been assigned a value. The arithmetic expression must then contain at least one **arithmetic operator**. The combination of values and operators in the expression allows a single value to be calculated and assigned to the variable on the left-hand side of the assignment statement.

For example, if a variable `myVariable` has been assigned a value 8, the expression:

```
myAnswer = 9 * myVariable
```

would result in a calculation of 9 × 8 to give the answer 72, which is then assigned to the variable `myAnswer`.

Java has many arithmetic operators. We will only consider the ones shown in Table 3.4.

Arithmetic operation	Symbol used in a Java program
Addition	+
Subtraction	−
Multiplication	*
Division	/
Modulus	%

Table 3.4: Arithmetic operators used in this chapter

QUICK QUESTION

What is the reason why one of these operations could give an unintended wrong answer if an expression were not formulated correctly?

There are two comments needed here about how Java is different from most other languages in its use of arithmetic operators.

1 In other languages (and in pseudocode) there can be two operators used specifically for handling integer division:

   • DIV for the whole number obtained from the division

   • MOD to give the remainder or **modulus** from the division

   Java uses the / symbol whatever type of values are used in the division and uses % as its version of MOD.

2 Most languages have a power symbol. For example 2^3 which has the value 8 would be represented in an expression by 2^3. Java does not include ^ as one of its operators. The Java approach will be explained in Section 7.7 of Chapter 7.

KEY WORD

modulus: an operator that produces the remainder if one integer is divided by another integer.

We can now consider a few examples of assignment statements in Java to see how they work. Consider that you have created the following variables with initialised values:

```
int number1 with value 75
int number2 with value 8
float number3 with value 8.0
int result1 with value 1
float result2 with value 1.0
```

These values will be used in the following examples.

For each example, you should first treat it as a task. Work out what you believe will be the value supplied to the variable on the left-hand side, then read the explanation that follows for the correct answer.

Example 1

```
number1 = number1 + 1;
```

There is no problem with having the same variable on each side of the statement. The value 75 is retrieved from memory and 1 is added, then the new value 76 is stored in the same memory location.

Example 2

```
result1 = number1 * 2 + number2;
```

This looks straightforward: 75 is multiplied by 2 to give 150 then 8 is added to get 158 stored for `result1`. However, it is not that simple as the next example will show.

Example 3

```
result1 = number2 + number1 * 2;
```

This gives the same answer! This is because an expression is not evaluated from left to right. Each operator has a **precedence**. In these two examples, the multiplication operator is evaluated before the addition operator, irrespective of how the expression is written.

If you wanted to force this expression to give the answer calculated as 'number2 added to number1, then this value doubled', you would have to write this as in Example 4.

KEY WORD

precedence: a property defined for each operator to direct the order in which an expression is evaluated.

Example 4

```
result1 = (number2 + number1) * 2;
```

This shows how parentheses can be used to control the order of evaluation of an expression. Here 8 is added to 75 to give 83 which is multiplied by 2 to give the value 166 for result1.

In these examples, we see how division and multiplication have the same precedence. If both are present, they are used in the order presented. Similarly, addition and subtraction have equal precedence. However, the multiplication and the division operations are performed before the addition or subtraction operations, irrespective of the order of the expression. Any expression in parentheses takes precedence over division and multiplication.

TIP

You may have learnt BODMAS or BIDMAS in your mathematics lessons. The same precedence exists in programming.

Example 5

```
result1 = number1/number2;
```

This gives the value 9 stored in result1. The reason the answer is 9 and not 9.375 is because both values in the division are integers. The evaluation uses integer division. Integer division calculates the largest whole number remaining from the division. The calculation ignores any remainder.

Example 6

```
result1 = number1 % number2;
```

This gives the value 3 stored in result1 representing the remainder from the integer division.

Example 7

```
result2 = number1/number3;
```

The value stored in result2 is 9.375. If either or both values in a division are not integers, then a normal division result is calculated.

You should note that in these examples, the value used to initialise result1 is never used. This will often happen. It is always good practice to initialise a variable with a sensible value early in a program even if you know that a new value will later be assigned. The reason is that the initialisation may help later in locating errors in a program.

ASSIGNMENT STATEMENTS CONTAINING ARITHMETIC EXPRESSIONS

This Skills Focus will illustrate what you will need to consider when choosing assignment statements to use in a program.

Let's consider that a program is to be used to calculate a reasonable estimate of the time it will take for a bus to be driven from one bus station to another. The requirements for this program are as follows:

- The journey has to involve stopping for a break at two other bus stations.

- The program must assume a value of 40 miles per hour for the average speed when the bus is travelling.

- The program must allow 45 minutes for each break.

- The program should add on a time to allow for unavoidable delays which are likely to happen from time to time. This will be calculated as 10% of the original estimate.

The best starting point is to consider what arithmetic will be needed.

The overall time will be calculated as the sum of:

- the time for each travelling part

- the time for each break

- the time added on to allow for unavoidable delay.

To calculate the time for each travelling part, we have to remember that speed = distance/time. This can be rearranged to see that time = distance/speed.

You need to remember that 10% can be calculated by dividing by 10.

You will need to make a decision about the units used for distance, time and speed. Sensibly time will be in hours so the 45 minutes for the break will become 0.75 hours.

We can now consider the variables we will need. All will be of type `float`. The following would be suitable:

```
distance1
distance2
distance3
timeForTravel
averageSpeed
timeForBreak
overallTime
```

The above list assumes that you are not going to need three separate variables for the calculation of the time for travel. You should note that `timeForBreak` is listed as a variable, but it would in practice be created as a constant value.

You can now decide how many assignment statements you will use. It would be possible to use just one but you may use more. Using more reduces the chances of errors and can make the code more understandable.

CONTINUED

Now is a good time to attempt to create a suitable set of assignment statements yourself before looking at the solution suggested below.

A possible solution is the following:

```
timeForTravel = (distance1 + distance2 + distance3)/averageSpeed;
timeForTravel = timeForTravel + 2 * timeForBreak;
timeForTravel = timeForTravel + timeForTravel / 10;
```

Questions

1   Can you see how these assignment statements might be combined into one assignment statement?

2   Would the assignment statements you created yourself combine to give the same result?

3   If you did choose to create just one assignment statement, what did you incorporate to make the logic clear?

Tasks

Chapter 4 contains details about how input and output are handled in Java and in pseudocode. For this reason, you are not required to include Java code for input or output in any task in this chapter. However, you will be able to add these features in answers to tasks set in Chapter 4.

If you are creating a pseudocode design for a task in this chapter, you can include simple code for input and output. The following are examples for variables `valueIn` and `valueOut`:

```
INPUT valueIn
OUTPUT valueOut
```

DEMO TASK 3.2

Consider a scenario where a group of neighbouring farmers buy seedlings in bulk then share the cost. A program is needed to calculate the amount each farmer has to pay assuming equal numbers of seedlings are bought by each farmer. There is a delivery charge of 10% of the cost of the seedlings to be included in the calculation.

Solution

A pseudocode design is:

```
INPUT numberOfFarmers
INPUT priceOfOneSeedling
INPUT totalNumberBought
totalCost ← totalNumberBought * priceOfOneSeedling
totalCost ← totalCost + totalCost * 0.1
farmerPriceToPay ← totalCost / numberOfFarmers
OUTPUT farmerPriceToPay
```

A flowchart version of the design is shown in Flowchart 3.1.

CONTINUED

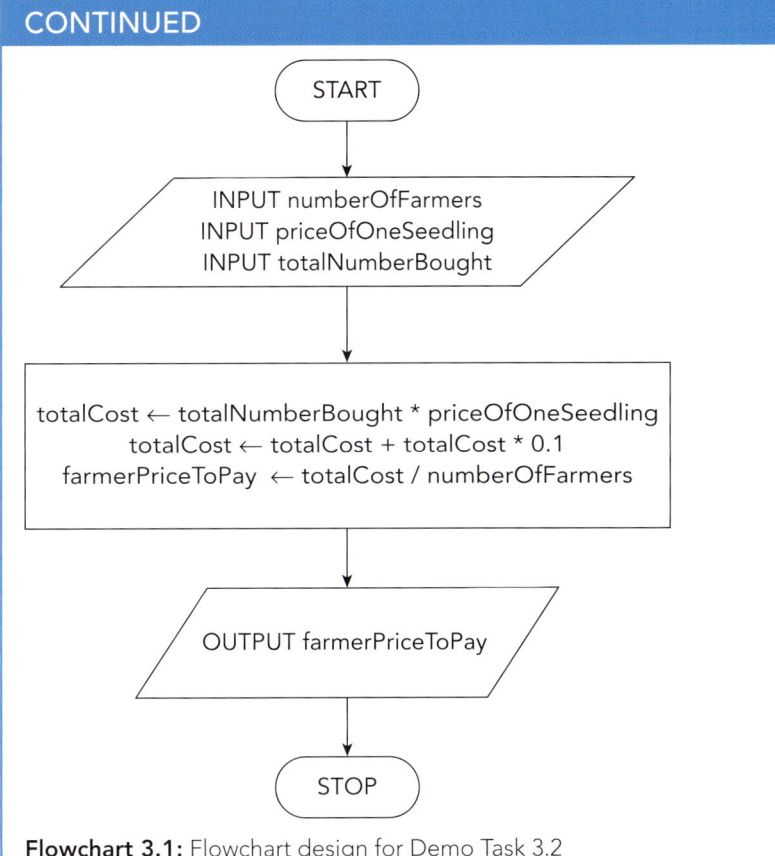

Flowchart 3.1: Flowchart design for Demo Task 3.2

The Java code for declarations and calculations could be:

```
int numberOfFarmers, totalNumberBought;
float priceOfOneSeedling, totalCost, farmerPriceToPay;
totalCost = totalNumberBought * priceOfOneSeedling;
totalCost = totalCost + totalCost / 10;
farmerPriceToPay = totalCost / numberOfFarmers;
```

PRACTICE TASKS 3.2–3.4

For each of the following tasks, you should provide a pseudocode or flowchart design that includes the input and output. Then Java code is needed only for the declaration of the variables and the calculation statements.

3.2 A multiply machine is required that handles the multiplication of two values. Therefore, the only input needed is of the two values to be multiplied.

CONTINUED

3.3 You need a program to calculate the volume of water needed to fill a small aquarium. The aquarium has rectangular sides and base. You know the values for the dimensions of the aquarium measured in centimetres. You need a calculation of the volume in litres.

3.4 A gardener creates lawns for clients by growing the grass from seed. She needs a program to calculate how much grass seed is needed for a particular lawn. The program also needs to calculate the cost of the seed.

The following facts are relevant here:

- 50 grams of grass seed are needed to cover an area of 1 square metre.

- The cost of grass seed is $20 per kilogram.

- The lawn will be rectangular in shape, and the client will supply the dimensions in metres.

TIP

1 litre is equal to 1000 cubic centimetres.

3.7 Use of an identifier table as part of the design process

Earlier in the chapter, we discussed how it is sensible to make decisions about the variables you need before you start to write any program code. The best way to record these decisions is to construct an identifier table.

DEMO TASK 3.3

You are considering writing a program to solve a problem introduced in Chapter 2. This is where a teacher wants a program that will calculate a percentage mark and a corresponding grade for a given exam for one candidate. The candidate has an identification number.

Your task is to create an identifier table for this program.

Solution

You decide the following:

1   Marks given are whole numbers; there are no half marks. Therefore, variables of type `int` will be needed to store marks.

2   In order to calculate a percentage mark, you need to know what the total mark is for the paper.

3   However, when the percentage is calculated, this will only occasionally be a whole number so it should be stored as type `float`.

4   You are never going to use the identification number in any numeric calculation. This is an example of a **pseudo number**. A pseudo number is always stored as type `String`.

KEY WORD

pseudo number: a collection of numeric digits that are never intended to be used in a numeric calculation; for example, a telephone number.

CONTINUED

5   Because there are only four grades, they can be recorded as single characters P, M, D and F, representing Pass, Merit, Distinction and Fail.

6   At this stage, you do not need to concern yourself with the criteria for determining the conversion of a percentage to a grade. You expect to write the code using literals so variables will not be needed.

7   However, you realise that you may have to reconsider this decision if the teacher says that the criteria are not always the same for different exams. In that case several more variables would be needed.

8   It makes sense for the program to output the exam name as well as the candidate grade and the candidate identification number. Because the same exam name will be used several times, this name must include the date of the exam.

You are now in a position to create the identifier table. This needs to record your chosen variable names with the data types, but it is also essential to include some information explaining the reasons for the choices. You decide that Table 3.5 would be a suitable identifier table for this problem.

Variable name	Variable type	How it will be used
candidateMark	int	This is the actual mark the candidate scored.
totalMarkPossible	int	This is the total mark available for the paper.
percentageMark	float	This is the candidate mark converted to a percentage.
candidateNumber	String	Each candidate has a unique number; an example of a value would be "45".
examName	String	This needs to include the date the exam was sat because the same exam name will be used more than once.
Grade	char	This is used to record fail, pass, merit or distinction using 'F', 'P', 'M', or 'D'.

Table 3.5: The identifier table for the teacher grading program

PRACTICE TASK 3.5

A company running a long-distance bus service needs a program to handle the issue of tickets. Part of this program will have to calculate the price a customer must pay. The customer might wish to purchase tickets for more than one person.

CONTINUED

The company policy for determining the price to pay includes the following features:

- Different prices will apply depending on which bus stations are to be used for the departure and for the arrival.

- For any combination of departure point and arrival point, there are different possible prices possible.

- The calculation of the price is always based on starting with the standard single journey fare for the particular bus journey.

- An individual ticket only covers one journey for one person.

- This standard single journey ticket price is then discounted depending on a number of possible factors; some examples of circumstances where a discounted price is to be calculated are shown in Table 3.6.

Circumstance	Ticket policy
An outward single ticket is needed plus a return single ticket for the same person.	There is a reduced (discounted) price for the return ticket.
A child less than 5 years old is travelling with a parent.	The child gets a free ticket for each journey.
A child aged 5 years or more but less than 10 years old is travelling with a parent.	There is a discounted price to pay for the ticket for the child for each journey.
A person aged 75 years or more is travelling.	The person gets a free ticket for each journey.
A total of at least six individual tickets are being paid for all together at least one month in advance of the first of any of the journeys needing tickets.	A discount is applied to the total amount to pay.

Table 3.6: Prices of tickets and discounts

a   Construct an identifier table containing a set of variable names that you would consider necessary if the ticket policy discounts are to be included in the calculation of ticket prices.

b   Write the declaration statements for these variables using Java code.

c   Write two assignment statements in Java that calculate the price of a booking for a mother and her nine-year-old son to travel to the shops (one way).

You do not need to use any values for the variables. Only variable names should be used.

CHALLENGE TASKS 3.1–3.4

3.1 Continuing with the bus company program, consider the following specific purchase of a number of tickets:

> The mother and the father of a child aged 7 are travelling with the child and with the child's grandmother aged 76. All four of them need tickets for an outward and a return journey. They are booking the tickets about six weeks ahead of the date for the outward journey.

Use Java code to construct a single assignment statement for the calculation of the price to pay for this collection of tickets.

You do not need to use any values for the variables. Only variable names should be used.

You may find it best to write a number of individual assignment statements first to see how these might be combined.

You can choose to tackle one or more of the following three tasks and supply all three of the requested deliverables for each one. Alternatively, you may wish to provide just part of what is requested.

For each task, provide the following:

- An identifier table.

- A pseudocode or flowchart design which includes input and output.

- Java code that includes variable declarations and the calculations.

3.2 A farmer who produces grain wants to calculate the number of bags he will need to take the grain to market. He also wishes to calculate how much he will be paid for the grain he has produced.

The following facts need to be considered for this program:

- One bag holds 25 kilograms of grain.

- He will charge $20 per bag.

- The calculation needs to allow for 5% of the grain to be spilt in the process of filling the bags.

- The calculation only needs to concern the number of full bags that will be used.

CONTINUED

3.3 You have been asked to deliver a program to the manager of the local swimming pool. The program needs to calculate the time that it will take to drain all of the water out of the pool when it has to be cleaned.

The manager will supply the values for the volume of the pool in litres and for the rate at which the pool can be drained in litres per minute.

The request is for the time to be output in hours and minutes.

3.4 Consider again the scenario provided for Demo Task 2.1 in Chapter 2. A program is needed to calculate the number of tins of paint that would be needed to paint a wall. A preliminary Structured English design was created in the Chapter 2 Demo Task 2.1.

The data needed for the program are:

- The length and the height of the wall.

- The area that can be covered with one tin of paint.

- The number of coats that need to be applied.

SUMMARY

A variable is a name associated with a memory location.
A declaration statement is used to provide a name and a data type for a variable.
The value assigned to a variable can be changed any number of times.
A constant can be declared with a value that can be used unchanged throughout a program.
It is important to initialise a variable before it is used in a program.
The left-hand side of an assignment statement contains only a variable name.
The right-hand side of an assignment statement contains an expression that must be evaluated to provide a new value for the variable on the left-hand side.
A value can be supplied to an expression as a variable, a constant or a literal.
Operators used in the evaluation of an expression have different precedences.
An identifier table can be constructed as part of the design process.

END-OF-CHAPTER QUESTIONS

1 Please refer back to Table 3.2.

   a Can you think of the reason why there must be a defined range for the values allowed for a variable of type `int`?

   b When would this defined range be most likely to cause a problem during execution of a program?

   c Do you know how a programming language might include a feature to minimise the possibility of such a problem occurring?

2 The following is a flowchart for the program discussed in Chapter 2 to convert an exam mark to a percentage.

Please supply a pseudocode and a Java code version of what would be required to match the process symbol in this flowchart. You will have to think carefully how you can handle the division operation. You can use some of the variable names from Table 3.5.

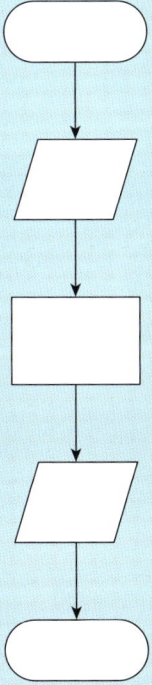

3 Consider the following problem definition:

A program is required that calculates the amount of money to be paid to an employee for a week's work. The calculation must include a tax deduction.

Construct an identifier table for this program. Include all variables that you believe would be needed for the program. There is no unique answer here.

4 A system is required to calculate the cost of a taxi journey. The cost is calculated by using the following data:

- A fixed cost of $5.00 is charged for the first 2 miles.
- $1.50 is then added for each additional mile travelled.

Provide a pseudocode design for this system.

> Chapter 4

Options when learning programming

IN THIS CHAPTER YOU WILL:

- learn how to compile and run a Java program

- learn about the features that Java provides for handling input and output

- learn some best practice rules for dealing with input and output

- write Java code for programs consisting only of a sequence logic construct

- learn about the options for the use of comments in a Java program.

Introduction

The aim of this chapter is to show you how you can run a simple Java program. You will be introduced to how your program code must be presented to the Java compiler before it can be executed. There is a focus on input and output. The pseudocode for input and output is discussed together with how a Java program can receive input and provide output. The chapter will also briefly discuss the following:

- The options for installation of Java.

- The options available for writing and running Java programs.

4.1 The Java Platform

You may wish to refer back to Chapter 1 at this point to remind yourself of the terminology used in the Java world. You need to have a version of the **Java Platform** installed before you can create and run Java programs. There are likely to be a number of different versions available. Because you will only be creating and running fairly simple programs, any available version should be suitable. We will assume that you have access to an installed version of Java in your school or college. If you have your own personal device, then there are several different versions of the Java Platform you can use. Because of the **portability** of Java, using more than one system will cause few if any problems.

A decision that you need to consider at some stage is whether or not to use an **Integrated Development Environment (IDE)**. This is because when the Java Platform is installed, it does not provide a built-in IDE. If you decide that you do wish to use an IDE you have to download and install this separately. Some of the options that are available are introduced later in this chapter. At this stage we will assume that you have not yet decided to use an IDE. We will show you how to run programs using the built-in facilities which are part of the Java Platform. Before moving on to Java, it is useful to consider some guiding principles regarding input and output.

4.2 Input and output

This section illustrates some general rules concerning input and output. Table 4.1 contains some examples of pseudocode design for input and output statements together with explanations.

Pseudocode	Explanation
OUTPUT answer	The value assigned to the variable answer is output.
OUTPUT answer1, answer2, answer3	There can be more than one output item; commas separate the items.
OUTPUT "Thank you "	There is output of the String literal.
OUTPUT "The answer is ", answer	A String literal can be combined with a value stored for a variable to make a meaningful output. Note the space at the end of the String literal.

(continued)

KEY WORDS

Java Platform: the product that contains the programming language plus supporting software.

portability: a measure of how easy it is to transfer software from one system or device to a different system or device and make it usable.

Integrated Development Environment (IDE): software that helps programmers to design, create and test program code.

Pseudocode	Explanation
INPUT firstNumber	This assigns the value that is input to the variable firstNumber. However, the user will not know what to input.
OUTPUT "Please enter a number " INPUT firstNumber	This outputs a prompt to the user so that the user knows what kind of input is expected.

Table 4.1: Examples of pseudocode design for input and output statements

You can see from these examples that there is no restriction on the number of values that are output. Input only brings in one value. It should be noted that the expectation when using the pseudocode OUTPUT and INPUT is that the output is to the computer screen and the input is from the keyboard.

4.3 Object-oriented programming (OOP)

Chapter 1 described how Java had its origins in the C and C++ programming languages. A major difference between C++ and C is that C++ has object-oriented programming (OOP) capability. This has carried forward into Java.

OOP is a powerful technique for approaching the development of complex programs. If you choose to progress further in your study of programming, you will definitely need to gain an understanding of OOP. At this stage, though, you do not need this understanding.

You cannot run a Java program without using some OOP features. However, you can treat the OOP features as a framework. You do not need to understand the components of this framework; you just need to know how to use it. You will see that running a program is essentially a process of filling in blanks inside the OOP framework.

The Appendix of this book does contain some explanation of the OOP features in Java. You might wish to look at that when you have successfully run a few programs.

4.4 Running a simple program using the Java Platform

This section is going to describe the stages of creating and running a program. The following program code will be used to illustrate what needs to be done. This code is for just about the simplest program you could possibly write.

```
class Hello {
public static void main(String args[]) {
    System.out.println("Hello");
}}
```

Code snippet 4.1

TIP

Remember that whenever you are writing a program that requires input from a user then you should always provide a prompt to inform the user making the input.

In Section 4.3, we talked about using a framework that we would write our code in. Let's look at what this means in the program code we have just seen. The normal black text is the OOP framework that needs to be included to make the program work. The blue text is our program.

```
class Hello {
public static void main(String args[]) {
    System.out.println("Hello");
}}
```

Some explanations of the code are given in Table 4.2.

Code	How it is used
`class Hello {`	`class` is part of the framework and must not be altered.
	`Hello` is a name that you have to choose to identify the class.
	`{` is a standard Java feature to identify the start of some lines of code.
`public static void main(String args[]){`	This is part of the framework and must not be altered.
	In particular, note that only `String` starts with an upper case letter.
`System.out.println("Hello");`	This is your one-line program which outputs Hello.
`}}`	The first of these `}` shows the end of your program code; the second `}` shows the end of the definition of the class named `Hello`. Both of these are part of the framework and must not be altered.

Table 4.2: Explanations of features in the simple Java program that outputs Hello

The program produces an output using a rather lengthy line of code that we do not need to discuss at this stage. Section 4.5 explains the input and output features in Java. For now, we can focus on how this program should be created and run.

Before you start, it is sensible to create a directory or folder specifically for storing Java programs. The next step is to create the code for the program using a **text editor**.

You now need to save the file containing this code in the directory or folder you have created for your Java programs. You do not have a choice for the filename. Because you have `class Hello {` in the program, this filename must be Hello.java. The Hello is exactly the name that you chose as a name for the class in the framework. It **must** have the same upper case first letter.

We now need to see how the Java Platform compiles and runs a program using simple commands typed in at the **command line prompt**.

KEY WORDS

text editor: software that allows you to create plain text with none of the stylistic features offered by a word processing application.

command line prompt: a facility provided by an operating system that allows keyboard input and screen output associated with running a program.

In preparation for using the Java commands, you must do two things:

1 Ensure that the command prompt is being used within the folder or directory you have created to store your Java program files.

2 Ensure that the commands provided by the Java Platform have been made available for use in this folder or directory.

How you do this will be dependent on the operating system you are using.

When everything is ready for action, you first have to compile your program. This needs the **javac command**. At the command line prompt, you type in:

```
javac Hello.java
```

If you have typed in your program correctly, this will just return you to the command line prompt. You will find that a new file called `Hello.class` has been created in the same directory or folder. This is the file containing the **Java Byte code**. If you have typed in your program incorrectly, you will receive an **error message** or, more usually, several error messages.

When no errors are present you can run the program using the **java command**. At the command line prompt, you type in:

```
java Hello
```

Note that you do not need to include any filename extension. You should now see the word Hello appear at the command line prompt in accordance with the code written for your simple program.

> ### PRACTICE TASK 4.1
>
> Create and run the Hello program that is listed in Code snippet 4.1 in Section 4.4.

4.5 Input, output and comments in Java code

You saw in Section 4.2 that the pseudocode used for input and output is straightforward. The Java coding for input and output is more complicated.

Coding will be easier if you create a text file with Java coding you use regularly. You can then copy and paste what you need when you are writing a new program. One component could be a template based on the framework introduced in Section 4.4. The idea of a template is that you have parts that you do not need to change and other parts where you have to supply appropriate code. The following is a suggestion for a template that you could include in the text file.

```
import java.util.Scanner;
**
class ** {
public static void main(String[] args) {
    **
}}
```

TIP

If you have never used the command line prompt before, now is the time to find out how this is accessed in the operating system you are using.

TIP

Take advice on how to ensure that the Java commands can be used in your Java program directory or folder.

KEY WORDS

javac command: the command you use at the command line prompt to compile a Java program.

Java Byte code: an intermediate code for a program that can be ported to different hardware devices.

error message: an output from the compiler caused by your code containing syntax errors.

java command: the command you use at the command line prompt to run the Java Byte code.

Each time that you wish to create a new program, you can simply replace each instance of ** with the relevant Java code. Table 4.3 explains the new features in this template:

Feature	Explanation
`import java.util.Scanner;`	`import` has to be included if you wish to use a facility provided by what Java calls a package (the Appendix has an explanation of this term).
	In this case you can now use `Scanner` in your program to provide input. Its use is shown in the program that follows this table.
	You might notice that this was not included for the very simple program introduced in Section 4.4.
`**`	The first example allows a comment to be included before the class statement.
`**`	The second must be replaced by a name.
`**`	The third has to be replaced by the program code.

Table 4.3: Explanations of some template features

KEY WORD

comment: a description of the algorithm written within the code. The comments are intended to help explain how the code works. Comments are ignored when the code is executed.

The following is the code for a small program which illustrates the use of input, output and comments. The different parts of the program have been colour-coded to help you to understand the contents by referring to Table 4.4.

```
import java.util.Scanner;
/*
to illustrate input of a string
 and of an integer
 */
class InputProg {
public static void main(String[]
args) {
int num,sum;
num = 10;
sum = 0;
System.out.println(
"What is your name");
Scanner nameInput = new
Scanner(System.in);
String yourName =
nameInput.nextLine();
System.out.println("Hello " +
yourName);
System.out.println(
"Enter a number between 1 and 10");
// the program will fail
// if an integer is not entered
Scanner numberInput = new
```

```
Scanner(System.in);
int yourNumber =
numberInput.nextInt();
sum = num + yourNumber;
System.out.println("Your number plus 10 =  " + sum);
}}
```

Code snippet 4.2

The code in Code snippet 4.2	Explanation

`import java.util.Scanner;`
`class       {`
`public static void main(String[] args) {`
 	
`}}`
	This is the framework that you had stored in your template.
	
	In other programs, the first line might be changed.
	
	The other lines must always be present.

`/*`
`to illustrate input of a string`
`and of an integer`
`*/`
	An example of a multi-line comment. It can contain any number of comment lines within `/*` and `*/`. (These are called 'delimiters'.) It is useful to include such a comment here to say what the program does.

`InputProg`	The name you have chosen. The file containing this code must be named `InputProg.java`.

`int num,sum;`
`num = 10;`
`sum = 0;`
	The declarations and initialisations at the start of your program.

`System.out.println(`
`"What is your name?");`
`Scanner nameInput = new Scanner(System.in);`
`String yourName = nameInput.nextLine();`
	The first example of a combination of three statements used to input a value.
	
	The first statement outputs some text to the screen as a prompt for the user
	
	The second and third statements use `Scanner`.
	
	Note that `nameInput` is used in the second and third statements.
	
	The third statement declares a variable `yourName`.
	
	`.nextLine` ensures that the input is used as a `String` value.

`System.out.println(`
`"Enter a number between 1 and 10");`
`Scanner numberInput = new Scanner(System.in);`
`int yourNumber = numberInput.nextInt();`
	The second example of three statements needed to input a value.
	
	The difference here is that an integer value is input. This is stored as a value for the variable `yourNumber`.
	
	`.nextInt` ensures that the input is used as an `int` value.

(continued)

The code in Code snippet 4.2	Explanation
`System.out.println("Hello " + yourName);`	The first of two examples where more than one item is output to the screen with one statement. Individual items are separated by + and the list of items is enclosed in parentheses.
`// the program will fail` `// if an integer is not entered`	You can use // at the beginning of a line to create a comment. Alternatively, you can use // at the end of a line of program code to attach a comment.
`sum = num + yourNumber;`	The one bit of arithmetic.

Table 4.4: Explanations of program code

There are some general points to make here:

1   Many of the lines of code have been continued on the next line. This was so that they would fit into the table. However, it does illustrate that this is possible because Java ignores any blank lines, blank spaces or indents that you use.

2   The approach to input and output used here is not the only one available for Java but it is suitable for when you are learning to program.

3   As always with Java, you must be careful to use upper or lower case, as shown in the examples.

4   To input a `real` value, the following could be used:

```
Scanner realInput = new Scanner(System.in);
float yourNumber = realInput.nextFloat();
```

5   There is no capability to input a `char` value using `Scanner`.

6   The examples of input contain a declaration of a variable. This should not be done if the program has earlier declared the variable. For example, if the variable `yourNumber` has previously been declared, the following line is used instead:

```
yourNumber = realInput.nextFloat();
```

4.6 Integrated Development Environments

As mentioned in Section 4.1, using an IDE is an option. There are many features that an IDE might provide. A program can be developed inside the environment. There can be features that assist in ensuring that correct syntax is used in the program code being written. When the program has been written, the program can be run inside the environment. The software handles the use of the compilation and run commands. There is no access to the operating system command line prompt. Input and output are handled through screens provided by the software.

If you do use an IDE, you still need to present a Java program using the framework that has been described in Section 4.5. You will still have the opportunity to use a text editor to write program code before you copy the code into the IDE so that it can be compiled and run.

If you are considering using an IDE the following are some of the options available at the time of writing:

- IntelliJ

- Eclipse

- NetBeans

- JEdit

- Dr Java

- JSource.

Of these, Eclipse or NetBeans are the ones most likely to be used by professional programmers. Others are more directly aimed at beginners.

4.7 Some tasks using arithmetic

Only the sequence logic construct is needed for these tasks.

DEMO TASK 4.1

You are asked to write a program that will input three separate numbers. The program will then calculate the sum of the three numbers and output the sum of the numbers and their average.

Solution

After giving this problem some thought, you decide that the numbers can have fractional parts; they will not be integers. You remind yourself that if you are summing values, the variable used for the sum must be initialised to zero before any values are added.

You decide that a pseudocode design will be helpful and that this should include variable declarations. In this you are going to use meaningful names for the variables. You create the following pseudocode.

```
DECLARE number1, number2, number3, sum, average : REAL
sum ← 0.0
OUTPUT "Enter first number"
INPUT number1
OUTPUT "Enter second number"
INPUT number2
OUTPUT "Enter third number"
INPUT number3
sum ← number1 + number2 + number3
average ← sum/3
OUTPUT "the sum is ", sum, " the average is ", average
```

You now wish to write the Java program. You realise that you cannot include a prompt in a statement performing the input. Therefore, you will need three statements for each of the three inputs. You decide to declare the variables used to store the values for the numbers in the input statements. This is because these numbers are not used in the program until after they have been input. However, the declarations for sum and average will be placed at the start of the program. Because you know it is good practice, you decide to initialise average at the beginning of the program as well as initialising sum. Your program (with colour-coding as used before) is:

```
import java.util.Scanner;
/*
```

CONTINUED

```
  calculates sum and average
  */
  class SumCalc {
  public static void main(String[] args) {
      double average = 0.0;
      double sum = 0.0;
      System.out.println("Enter first number");
      Scanner mynum1 = new Scanner(System.in);
      double number1 = mynum1.nextFloat();


      System.out.println("Enter second number");
      Scanner mynum2 = new Scanner(System.in);
      double number2 = mynum2.nextFloat();


      System.out.println("Enter third number");
      Scanner mynum3 = new Scanner(System.in);
      double number3 = mynum3.nextFloat();
      sum = number1 + number2 + number3;
      average = sum/3;
      System.out.println("Sum is " + sum + " Average is " + average);
      }}
```

Some explanations of the program coding are given in Table 4.5.

Code	Explanation

`import java.util.Scanner;`
 	
`/*`
`calculates sum and average`
`*/`
 	
`class SumCalc {`
`public static void main(String[] args) {`
	The framework as before with the insertion of `SumCalc` as your choice of name for the class and your choice of comment inserted.

`double average = 0.0;` `double sum = 0.0;`	Declarations and initialisations of two variables. The use of the data type `double` is explained after the table.

`System.out.println("Enter first number");`
`Scanner mynum1 = new Scanner(System.in);`
`double number1 = mynum1.nextFloat();`
 	
`System.out.println("Enter second number");`
`Scanner mynum2 = new Scanner(System.in);`
`double number2 = mynum2.nextFloat();`
 	
`System.out.println("Enter third number");`
`Scanner mynum3 = new Scanner(System.in);`
`double number3 = mynum3.nextFloat();`
	Three sets of inputs of values of type `double`.
	
	Note that the code can still use `.nextFloat`.

(continued)

CONTINUED

Code	Explanation
```sum = number1 + number2 + number3;``` ```average = sum/3;``` ```System.out.println("Sum is " + sum +``` ```"Average is " + average);```	The arithmetic and the output with helpful text included.

Table 4.5: Some explanations of Java code

You should have been expecting the code to be written using the `float` datatype. Unfortunately, there can be problems when this data type is used in a Java program. The `double` datatype is an alternative for variables storing real values. Practice Task 4.2 investigates this potential problem and solutions available.

PRACTICE TASKS 4.2–4.6

4.2 The aim of this task is for you to investigate the use of the data types `float` and `double`. There are three similar programs. You need to run all three programs.

The first program uses the `float` data type. The second also uses this data type but the coding uses `0.0f` to force the literal value to match a float value. The third uses the `double` data type.

```
float sum = 0.0;
System.out.println("Please enter a number");
Scanner mynum = new Scanner(System.in);
float number1 = mynum.nextFloat();
sum = 10.0 + number1;
System.out.println("The sum of your number and 10
    is " + sum);

float sum = 0.0f;
System.out.println("Please enter a number");
Scanner mynum = new Scanner(System.in);
float number1 = mynum.nextFloat();
sum = 10.0f + number1;

System.out.println("The sum of your number and 10
    is " + sum

double sum = 0.0;
System.out.println("Please enter a number");
Scanner mynum = new Scanner(System.in);
double number1 = mynum.nextFloat();
sum = 10.0 + number1;
System.out.println("The sum of your number and 10
    is " + sum
```

TIP

If you want to force a literal value to match a `float` value, place `f` at the end; for example, `25.0f`

CONTINUED

If you do not get an error message when you try to run the first example, it is likely that you are using an old version of Java. The expectation is that you will get an error message. This is a rare example of where simple programming has been affected by changes in version. The problem arises from the storage of the literal value, which is being assigned to the variable sum.

The other two programs should both run without any problem. In future programming tasks, please feel free to choose which version you will use. As with several aspects of using Java, for now, you just need to know how to get a program to work. Understanding fully the choice of techniques you are using can come later.

4.3 Create and run the program listed in Code snippet 4.2 in Section 4.5 that has the class named as InputProg. Use simple values for the numbers input so that you can check that the program is giving the correct output.

You created Java code for Practice Tasks 3.2, 3.3 and 3.4 in Chapter 3. It is now time to complete the programs by including input and output.

4.4 Complete the program for the multiply machine that inputs two numbers and outputs the value obtained by multiplying the two numbers together.

4.5 Complete the program that calculates the volume of water needed to fill an aquarium. The program needs to input the dimensions of the aquarium and output the volume.

4.6 Complete the program that calculates the amount of grass seed needed to create a lawn and the cost of that amount of seed. The calculation uses the dimensions of the lawn and the following facts:

- 50 grams of grass seed are needed to cover an area of 1 square metre.

- The cost of grass seed is $20 per kilogram.

You will need to decide if any values can be stored as constant values.

CHALLENGE TASKS 4.1–4.3

Chapter 3 contained three tasks (Challenge Tasks 3.2, 3.3 and 3.4) requesting Java code. If you did write the code, you can now complete the programs by including input and output. Alternatively, if you chose not to write any code, you now have the opportunity to write the code for the whole program.

4.1 Complete the program for the calculation of the number of bags a farmer could fill from the grain harvest. The program is to input the total amount of grain and the selling price for a bag and output the number of bags and the amount of money the farmer receives for these bags.

4.2 Complete the program which calculates the time taken to drain a swimming pool. The program inputs the volume of the pool and the drainage rate and outputs the time in hours and minutes.

4.3 Complete the program that calculates the number of tins needed to paint a wall. The program needs input of the dimensions of the wall, the coverage provided by one tin and the number of coats needed. The program is to output the number of tins required to guarantee complete coverage of the wall.

SUMMARY

The Java Platform provides the commands javac and java for compiling and running a program at the command line prompt.
An Integrated Development Environment is an option for writing and running a program.
Program code has to be enclosed in object-oriented features for the code to be compiled and run.
A name must be given in a `class` statement that matches the name given to the file containing the code.
There are different codings for multi-line comments and for comments on a single line.
You can output more than one item in one statement.
You have to input items one at a time.
Using `Scanner` allows you to input a string, integer or real value.

END-OF-CHAPTER QUESTIONS

1   State what is best practice when writing code to provide input to a program. Give an example of Java code for the input of an integer value.

2   The following code has two errors. Identify what they are.

```
public static void main(String[] args) {
  int number1 = 0;
  int number2 = 0;
      int answer1 = 0;
      System.out.println("Please enter first number");
      Scanner myObj1 = new Scanner(System.in);
      number1 = myObj1.nextInt();
      System.out.println("Please enter second number");
      Scanner myObj2 = new Scanner(System.in);
      number2 = myObj1.nextInt();
      answer1 = number1 + number2);
      System.out.println(answer);
      }}
```

3   State which simple data type cannot be input using `Scanner`. Describe a simple method of coding that could solve this problem.

4   Consider the scenario from Demo Task 3.2 in Chapter 3, where a group of farmers buy a product in bulk and then share the product out. Write a Java program that calculates the price each farmer has to pay assuming each farmer buys the same amount of the product. The calculation must allow for the delivery charge that has to be paid when the product is first delivered. This delivery charge is calculated as 10% of the total cost of the bulk purchase.

> Chapter 5
Selection

Introduction

Chapter 2 introduced selection as a type of logic construct that could be used in an algorithm. A selection construct is needed when an algorithm has options for which action should be performed. Chapter 2 also introduced the diamond-shaped decision symbol that is used in a flowchart to define a condition to be tested to determine which action follows.

This chapter introduces the different types of selection construct that can be used in an algorithm. It also introduces the syntax used to define them in Java.

5.1 The basic concepts

A **selection construct** must define one or more actions that could be performed. It must also define the conditions for decisions to be made as to which actions should be chosen.

In everyday life, we regularly make decisions. These decisions are often dependent on whether the answer to a question is yes or no. For example, if the question were 'Is it raining?' and the answer were yes, then the action could be to take an umbrella. However, this approach is not used in an algorithm designed to be implemented as a computer program. Instead the approach is based on the use of **Boolean logic**. Everything in Boolean logic has one of the two possible values true or false. To use Boolean logic the question is now replaced by what we can call a **logic proposition**. In the example here, the logic proposition would be 'It is raining'. This looks like a statement of fact but it is not used as such. Instead it is used as a condition that can be tested. Then the action taken depends on whether this proposition is in fact true or false. Flowchart 5.1 shows a flowchart representation of this Boolean logic.

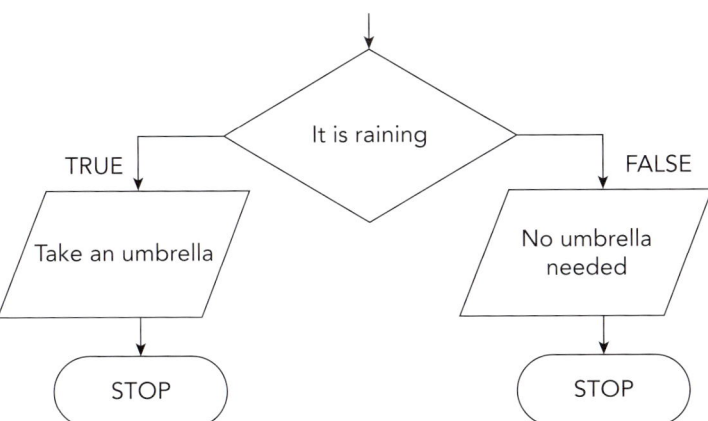

Flowchart 5.1: Flowchart showing a decision based on a logic proposition 'It is raining'

When you write a program that needs a selection construct you use an appropriate **conditional statement** (sometimes called a selection statement). Every programming language syntax includes the format for the different sorts of conditional statement that the language supports. Whatever the form of a conditional statement, it will define a test of a logic proposition (the statement of fact) and one or more actions that will happen if the logic proposition is true. There may also be actions defined to be performed if the logic proposition is false. Before we look at examples of conditional statements, we need to discuss the alternatives available for defining a logic proposition.

5.2 Relational expressions

One way of defining a logic proposition is to use a **relational expression**. A relational expression contains a comparison between two values using a **relational operator** (sometimes called a comparison operator). At least one of the values in a relational expression must be defined as a variable. A simple example is:

```
candidateMark > 75
```

There are two points to note here:

- The variables or literals (a value that you use directly in your program) used in a relational expression can be of any data type that allows a sensible comparison.

- The complete expression is a `boolean` data type with a value of either true or false.

Table 5.1 shows the relational operators that are used for pseudocode and the equivalent ones used in Java.

Pseudocode symbol	Java symbol	Meaning
=	==	is equal to
<>	!=	is not equal to
>	>	is greater than
<	<	is less than
>=	>=	is greater than or equal to
<=	<=	is less than or equal to

Table 5.1: Relational operators

KEY WORDS

conditional statement: coding used to program a selection construct. This statement contains a logic proposition and the definitions of dependent actions.

relational expression: a form of expression that contains a comparison of two values using a relational operator.

relational operator: an operator that is used to compare values. It can test whether values are equal or whether one value is greater than another value. (It is sometimes called a comparison operator.)

TIP

Mathematicians often use a combined symbol such as ≤ when writing a comparison. In Java you must use the two separate symbols <= as shown in Table 5.1.

QUICK QUESTION

Can you think of a reason why Java uses the symbol with two == signs rather than an ordinary = sign?

5.3 Logic expressions

'It is raining' as a logic proposition is unsuitable for use in a program. It is too informal. We will now introduce the term **logic expression** as an all-inclusive description of how you can present a logic proposition in a formal way for use in program code. There are three possible options:

1   The logic expression consists of one variable name. The variable must have been declared as data type `boolean`.

2   The logic expression consists of one relational expression.

3   The logic expression consists of variable names or relational expressions or both combined with at least one **Boolean operator**.

Table 5.2 contains definitions of the three Boolean operators that we need to consider in this book together with how they would be represented in pseudocode or Java.

Generic name	In pseudocode	In Java
AND	AND	`&&`
OR	OR	`||`
NOT	NOT	`!`

Table 5.2: Boolean operators

We can now look at three simple examples of logic expressions written in Java. In Chapter 3, Practice Task 3.5 and Challenge Task 3.1 both considered the scenario of a bus company needing a program to calculate the price to charge for tickets. The tasks in Chapter 3 only considered possible assignment statements using arithmetic expressions. However, there were a variety of conditions relating to the possibility of discounted prices. It is likely, therefore, that a complete program would also contain conditional statements.

The following are examples of logic expressions that could be used in such a program.

1   In this first example, `junior` (a variable name we can give the child aged 5–10) and `dayReturn` will have been declared as type `boolean`.

`junior && dayReturn`

2   In this second example:

`age < 5 || age >= 75`

there are two points to note:

•   Age 75 is included but age 5 is not included.

•   `age` has to be included twice because the following sort of construct is not allowed in Java:

`age < 5 || >= 75`

3   In this third example:

`student || age >= 75`

you should be able to see that the expression would only make sense if the relational expression with the age comparison were handled first. This will in fact happen because Java defines rules for precedence. One aspect of these rules is that relational operators are higher precedence than Boolean operators. In this

TIP

This is a good time to remind you that in Java the data type is `boolean` with no upper case B.

KEY WORDS

logic expression: a formal version of a logic proposition that contains a combination of Boolean values and Boolean operators that equates to an overall value of true or false.

Boolean operator: one of AND, OR or NOT (there are others but they will not be considered in this book).

example, the relational operator >= has higher precedence so is evaluated before the Boolean operator ||.

One particularly important example of precedence is the rule that:

NOT is higher precedence than AND, which is higher precedence than OR.

In the discussion in Section 3.6 of Chapter 3 concerning arithmetic expressions, it was pointed out that parentheses could be used to make the logic clear. You can do the same when writing a logic expression.

To illustrate the effect of precedence and the use of parentheses, we can consider the following three expressions where A and B are variables of type `boolean`.

```
!A && B
B && !A
!(A && B)
```

The first two of these are evaluated in the same way with the ! operator being applied first. For the expression to be evaluated as true, A must be false and B must be true.

In the third example, the parentheses cause the && operator to be used first. The ! operator is then applied. For the whole expression to be true, A && B must be false. This can happen in one of three ways:

- If A is false and B is true (as in the first two examples).

- If B is false and A is true.

- If both A and B are false.

> **TIP**
>
> If you decide to use parentheses in an expression, be sure to check that these are paired properly. It is easy to forget to close a parenthesis when you are using more than one pair.

SKILLS FOCUS 5.1

LOGIC EXPRESSIONS DEFINING A RANGE OF VALUES

We are going to look in a bit more detail at problems where we need to identify a range of values to determine a specific action. As an example, we can consider the condition defined in a program design discussed in Section 2.3 Chapter 2 where the teacher was assigning grades to exam results:

Grade as Merit if mark in range 61–80

We are going to look at some examples of Java coding for checking if a value falls in this range. Before any code can be written, we must choose a name for the variable that will be used to handle the exam mark. The range 61–80 must then be defined in a logic expression using this variable name. There are a number of points to make about the options for this:

1   The range can only be defined in a program by stating two conditions: one for the lower limit and one for the upper limit.

2   The simplest definition of the range is one that does not use >= or <=. This simplest version is then:

```
candidateMark > 60 && candidateMark < 81
```

It is vitally important that the correct values are used and that they are joined with the AND operator.

CONTINUED

3   There will always be alternative ways to define the expression. For example:

```
candidateMark >= 61 && candidateMark <= 80
```

4   You could define the range that is not allowed then to exclude it, as shown here:

```
!(candidateMark < 61 || candidateMark > 80)
```

Unfortunately, it is easy to make mistakes in writing these logic expressions. Most of your mistakes will be in the choice of the combination of the value and the relational operator. These mistakes will result in the range of values being incorrectly defined.

However, it is possible that you might write a logic expression that defies logic. For example:

```
candidateMark < 61 && candidateMark > 80
```

You can see that, whatever you intended at the time, this is nonsense. Unfortunately, the Java compiler will not notice the nonsense. It will assign the value FALSE to this expression and continue with compiling the program. It should be clear from this that it is always important that you double-check any logic expression you have just written.

Questions

You can assume that the following condition still applies:

Grade as Merit if mark in range 61–80

For each of the following examples, state whether the logic does define a range or is nonsense. Then if a range is defined, say whether or not it is the correct one.

1   `candidateMark > 60 && candidateMark <= 81`
2   `candidateMark < 81 && candidateMark > 60`
3   `candidateMark > 60 && candidateMark ! < 81`
4   `candidateMark > 60 || candidateMark < 81`

DEMO TASK 5.1

You are requested to create a program to calculate the price to be charged for use of a ferry. This is a rather large ferry that can carry vehicles as well as foot passengers (people). You have been given the following information:

- *A vehicle is charged differently depending on whether the vehicle is classified as a car or a van (where a van is any commercial vehicle).*

- *A person is charged differently depending on whether the person is classified as an adult or a child.*

- *Different charges are made for car passengers and foot passengers.*

CONTINUED

Solution

You give the problem some thought. There are four aspects that are unclear. After asking some questions, you get the following clarification:

- Commercial vehicles are not allowed to carry passengers.

- The charge for a vehicle includes the charge for the driver.

- A child is charged the same price whether in a car or on foot, and if there is a child there must be an adult.

- There are different charges for an adult depending on whether the adult is on foot or in a car.

- The price will be calculated for just a van or for just one car plus passengers or for just a group of foot passengers.

You now decide to create an identifier table. You intend to tackle the problem by using several `boolean` variables. The identifier table you create is shown as Table 5.3.

Variable name	Variable type	How it will be used
`vehicle`	STRING	To be used for input only.
`numberAdult` `numberChild`	INTEGER	Can be zero upwards; to be used whether the person is on foot or in a car.

`carPrice`
`vanPrice`
`carAdultPrice`
`childPrice`
`footAdultPrice`
`totalToPay`
	REAL	To be used in the calculation of the total price subject to decisions being tested to see if the charge applies.

`car`
`van`
`adult`
`child`
	BOOLEAN	All to be initialised to false at the beginning of the program.

`noAdultProblem`	BOOLEAN	To be initialised to false at the beginning of the program then to be assigned a value using the expression: `NOT adult AND child`

Table 5.3: Identifier table for ferry charges

You are going to learn about IF statements before considering a design for the program.

PRACTICE TASK 5.1

We return to the problem mentioned earlier in this chapter which was first introduced in Practice Task 3.5 in Chapter 3. The program has to calculate ticket prices of a bus journey. You are at an early stage of considering a design. Specifically, you are considering logic expressions that might be needed.

The task considers three examples of the rules that could apply. For each of the rules, the task requires you to provide Java code for:

a   any declaration statements that you need to include in the program before the logic expression could be used

b   the logic expression that would be suitable as part of the code defining the rule.

As an example:

Rule: A ticket is free for someone at least 75 years old.

You could then provide the following code fragment answers relating to this rule:

a   `int age;`

b   `age >= 75`

Note that you are not being asked to provide any code to handle the tickets being free. You are just being asked for code fragments relating to each of the following rules:

Rule 1: A child younger than 5 years old accompanied by a parent travels for free.

Rule 2: A student gets a concession if carrying a student card.

Rule 3: A season ticket holder does not need to have a ticket issued provided that the period for which the season ticket is valid has not expired.

5.4 IF statements

An **IF statement** is one of the two alternatives available for use as a conditional statement. It is the one that you are most likely to choose.

The simplest form of IF statement is where there is one condition tested and one action defined. As an example, if an exam mark, expressed as a percentage, were to be assessed simply as either pass or fail, the following might be included in the Java program:

```
if (candidateMark >= 40) grade = "Pass";
```

It is important that you note that the `if` is written in lower case and that the logic expression is written inside parentheses `()`. The action just consists of a normal assignment statement.

KEY WORD

IF statement: a statement that allows a program to follow or ignore a sequence of code depending on a Boolean condition.

In principle, you could include a statement like this in a program without including any action to be taken if the condition were not true. However, such cases will be rare exceptions. In this example, we do need to set the grade if the candidate has failed. The question is: how do we do this?

Flowchart 5.2 shows a fragment of a flowchart, which is a design for one solution. In this approach, the variable grade is initialised with a value 'Fail'. This remains the value unless the conditional statement changes it to 'Pass'.

An alternative flowchart design is shown in Flowchart 5.3.

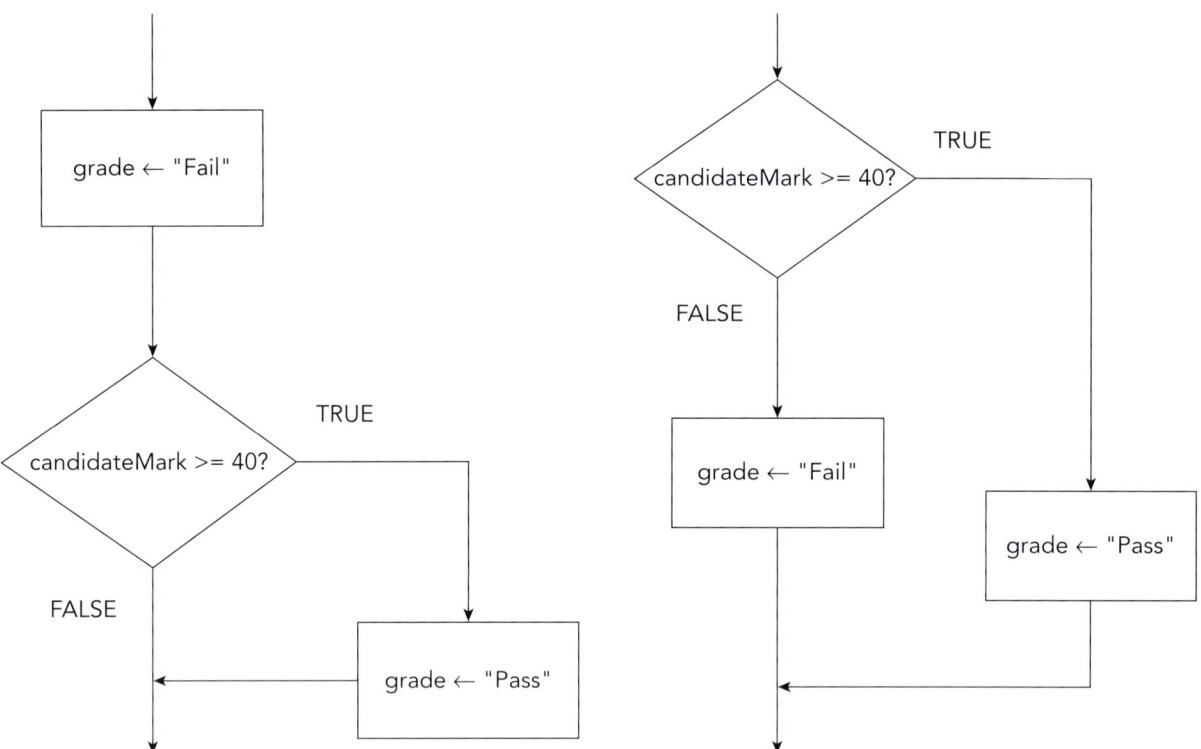

Flowchart 5.2: Flowchart example to grade exam mark   **Flowchart 5.3:** Alternative solution to grading exam marks

At first glance, Flowchart 5.3 looks to be straightforward. However, it doesn't give any information about the coding needed to ensure that the correct value is assigned to grade.

The solution is for a programming language to support a slightly extended version of an IF statement that includes an **ELSE** component (ELSE provides an alternative action). This can be illustrated by showing how the two options would be presented in pseudocode.

For the simple IF statement corresponding to Flowchart 5.2, the pseudocode would be:

```
IF candidateMark >= 40
  THEN
    grade ← "Pass"
ENDIF
```

KEY WORD

ELSE statement: a continuation following an IF statement which defines an alternative action.

The pseudocode corresponding to Flowchart 5.3 would be:

```
IF candidateMark >= 40
  THEN
     grade ← "Pass"
  ELSE
     grade ← "Fail"
ENDIF
```

The example here shows only one statement for each option. However, there can be an unlimited number of statements following the THEN (in the first example) and following the ELSE (in the second example).

The statements are not followed in a simple sequence. Either the statements after THEN are ignored or the statements after ELSE are ignored. The use of indentation is now really useful in making the logic of the code understandable.

Java program code has the same sort of structure but THEN and ENDIF are not included in the code. The following two examples show two alternatives for how the Java code could be written with the else component.

> **TIP**
>
> You will create code that is easier to understand if you include indentation. This is particularly helpful where the code has a number of statements that are not followed in a simple sequence.

Example 1

```
if (candidateMark >= 40) grade = "Pass";
    else
        grade = "Fail";
```

Example 2

```
if (candidateMark >= 40)
{
        grade = "Pass";
}

else
{
        grade = "Fail";
}
```

There are several aspects of Java that are illustrated by these codes:

- All keywords such as `if` and `else` must be written in lower case.

- The code can always use indentation and blank lines. These are for the benefit of anyone reading the code. Java just ignores such features.

- Although there is only one action for `if` and one for `else`, the code in Example 2 has been written to show how multiple actions can be included. It is a general feature of Java that multiple statements can be actioned by including them in a pair of braces (curly brackets). Each action in the braces must be terminated with a semi-colon.

PRACTICE TASK 5.2

Write the code in Java for the program design that has already been introduced in Section 5.4. The teacher inputs an exam mark as a percentage. The program must output either pass or fail. You can choose to base the program on Flowchart 5.2 or Flowchart 5.3. Run the program to check that it outputs the correct result.

Usually, a problem scenario will be such that there are several actions possible, with each action corresponding to a specific condition being true. There are several options for dealing with this type of scenario using IF statements. We need now to consider the options available.

SKILLS FOCUS 5.2

IF STATEMENTS WITH MULTIPLE CONDITIONS

To consider the different possibilities for using IF statements, we can consider the exam grade calculation introduced in Chapter 2. We can assume that marks have been converted to percentages. The rules for awarding grades are defined as:

Grade as Distinction if percentage above 80.

Grade as Merit if percentage in range 61–80.

Grade as Pass if percentage in range 40–60.

Grade as Fail if percentage less than 40.

With the knowledge you have so far, you would be likely to provide the following as a pseudocode design for coding these four rules:

```
IF candidateMark > 80
  THEN
    grade ← "Distinction"
ENDIF
IF candidateMark > 60 AND candidateMark < 81
  THEN
    grade ← "Merit"
ENDIF
IF candidateMark > 39 AND candidateMark < 61
  THEN
    grade ← "Pass"
ENDIF
IF candidateMark < 40
  THEN
    grade ← "Fail"
ENDIF
```

This code simply consists of four separate simple IF statements. This has the advantage that it is, in effect, a direct translation of the conditions defined and it is easy for someone reading the code to understand the logic. However, you need to be familiar with other options that are available which can provide better solutions.

CONTINUED

Option 1

This is where you keep the four different conditions but use ELSE IF options inside one overall IF statement. This can be described as a **nested selection statement** solution. The pseudocode is now as shown here.

```
IF candidateMark > 80
  THEN
     grade ← "Distinction"
  ELSE
     IF candidateMark > 60 AND candidateMark < 81
       THEN
          grade ← "Merit"
     ENDIF
  ELSE
     IF candidateMark > 39 AND candidateMark < 61
       THEN
          grade ← "Pass"
     ENDIF
  ELSE
     IF candidateMark < 40
       THEN
          grade ← "Fail"
     ENDIF
  ENDIF
```

Option 2

At this stage, you might be thinking that you do not need to include four separate conditions. Once you have included outcomes for three of the conditions, the remaining condition does not need to be included. Let's suppose that you decide to exclude the condition with candidateMark < 40.

One way of doing this is to use the code from Option 1 but replace the last ELSE IF with ELSE so the last part of the code becomes:

```
IF candidateMark > 39 AND candidateMark < 61
  THEN
     grade ← "Pass"
  ELSE
     grade ← "Fail"
ENDIF
```

The other way is to initialise the variable grade to have the value 'Fail' before the IF statement begins. Then the whole of the last ENDIF component in the Option 1 code can be removed, as shown here:

```
grade ← "Fail"
IF candidateMark > 80
  THEN
     grade ← "Distinction"
  ELSE
```

> **KEY WORD**
>
> **nested selection statement:** a coding structure where an IF statement is included inside another.

CONTINUED

```
        IF candidateMark > 60 AND candidateMark < 81
          THEN
              grade ← "Merit"
        ENDIF
      ELSE
        IF candidateMark > 39 AND candidateMark < 61
          THEN
              grade ← "Pass"
        ENDIF
  ENDIF
```

Option 3

It is possible to create variables with values corresponding to a logic expression. Using the Merit grade as an example, you would create a variable of type `boolean` and assign the matching logic expression as a value:

```
merit ← candidateMark > 60 AND candidateMark < 81
```

then inside the IF statement, you would have:

```
      ELSE
        IF merit
          THEN
              grade ← "Merit"
```

This approach would be particularly useful if a program contained more than one use of the logic expression.

Option 4

The final option is one that is available when logic expressions contain more than one relational expression. Instead of one combined condition, there can be two IF statements. This is a slightly different example of the use of a nested selection statement.

For example, if we consider the pseudocode:

```
      ELSE
        IF candidateMark > 60 AND candidateMark < 81
          THEN
              grade ← "Merit"
```

The following is the equivalent if a nested IF is used:

```
      ELSE
        IF candidateMark > 60
          THEN
              IF candidateMark < 81
                THEN
                    grade ← "Merit"
```

CONTINUED

Questions

1   Why do you think that using the ELSE IF code, as shown in Option 1, is preferable to using the four separate simple IF statements at the start of the Skills Focus?

2   Why do you think that using a nested IF in Option 4 is preferable to the code with two relational expressions connected with AND?

3   Could you do something similar if the code had two relational expressions connected with OR? For example:

```
candidateMark < 40 OR candidateMark >= 80
```

4   Write the corresponding Java code for the Option 4 pseudocode.

DEMO TASK 5.2

This is a continuation from Demo Task 5.1: the price to charge for passengers on a ferry. You want to create a pseudocode design and a flowchart design. You decide to include declaration statements in the pseudocode. You also decide that all of the charges can be declared as constants.

Solution

The pseudocode design for the declarations and initialisation is:

```
CONSTANT carPrice = 25.0 // includes the driver
CONSTANT vanPrice = 35.0 // includes the driver
CONSTANT carAdultPrice = 5.0 // for a passenger in a car
CONSTANT childPrice = 2.0
CONSTANT footAdultPrice = 3.0
    DECLARE adult, child, van, car, noAdultProblem : BOOLEAN
    DECLARE vehicle : STRING
    DECLARE numberAdult, numberChild : INTEGER
    DECLARE totalToPay : REAL
        totalToPay ← 0.0
        adult ← FALSE
        child ← FALSE
        van ← FALSE
        car ← FALSE
        noAdultProblem ← FALSE
        numberAdult ← 0
        numberChild ← 0
```

It is sensible to begin by checking if a van is to be paid for because there will then be no charge for passengers.

```
OUTPUT "enter vehicle type or none if foot passengers"
INPUT vehicle
van ← vehicle = "van"
IF van
   THEN
      totalToPay ← vanPrice
   ELSE
```

The remainder of the design has to input the passenger numbers, check there is an adult with children then calculate and output the price to pay.

CONTINUED

```
        OUTPUT "how many adults"
        INPUT numberAdult
// a driver would not be included in this number
        adult ← numberAdult > 0
        OUTPUT " how many children"
        INPUT numberChild
        child ← numberChild > 0
        noAdultProblem ← NOT adult AND child
        IF noAdultProblem
          THEN
            OUTPUT "no children without an adult"
        ENDIF
        car ← vehicle = "car"
        IF NOT van AND NOT noAdultProblem
          THEN
            totalToPay ← childPrice * numberChild
        ENDIF
        IF car
          THEN
            totalToPay ← totalToPay + carPrice +
    carAdultPrice * numberAdult
          ELSE
            totalToPay ← totalToPay +
    footAdultPrice * numberAdult
        ENDIF
        IF NOT noAdultProblem
          THEN
            OUTPUT totalPriceToCharge
        ENDIF
```

The flowchart for this design is shown in Flowchart 5.4. To make the diagram a little less cluttered, abbreviations have been used in places for variable names (see Table 5.4).

Abbreviation in flowchart	Variable name in pseudocode
carP	carPrice
cAP	carAdultPrice
nAdult	numberAdult
tot	totalToPay
vanP	vanPrice
childP	childPrice
fAP	footAdultPrice
nChild	numberChild

Table 5.4: Abbreviations for the flowchart

CONTINUED

Flowchart 5.4: Flowchart for ferry charges

A possible Java code would be:

```java
import java.util.Scanner;
/*
the ferry program
 */
class Ferry {
public static void main(String[] args) {
    final float carPrice = 25.0f;
    // note the addition of f to allow the use of float
    final float vanPrice = 35.0f;
    final float carAdultPrice = 5.0f;
    final float childPrice = 2.0f;
    final float footAdultPrice = 3.0f;
    boolean adult = false, child = false, van = false, car = false,
        noAdultProblem = false;
    String vehicle;
    int numberAdult = 0, numberChild = 0;
    float totalToPay = 0.0f;
```

CONTINUED

```java
        float totalToPay = 0.0f;
        System.out.println("Enter car, van or foot");
        Scanner vehicleInput = new Scanner(System.in);
        vehicle = vehicleInput.nextLine();
        van = vehicle.equals("van");// see comment below
        if (van)
            totalToPay = vanPrice;
        else
            {
            System.out.println("How many adults?");
            Scanner adultInput = new Scanner(System.in);
            numberAdult = adultInput.nextInt();
            adult = numberAdult > 0;
            System.out.println("How many children?");
            Scanner childInput = new Scanner(System.in);
            numberChild = childInput.nextInt();
            child = numberChild > 0;
            noAdultProblem = !adult && child;
            if (noAdultProblem)
                System.out.println("No children without an adult");
            }
            car = vehicle.equals("car");
            if (!van && !noAdultProblem)
            {
              totalToPay = childPrice * numberChild;
              if (car)
              totalToPay = totalToPay + carPrice +  carAdultPrice *
              numberAdult;
              else
              totalToPay = totalToPay + footAdultPrice * numberAdult;
            }
        if (!noAdultProblem)
            System.out.println("Total charge is " + totalToPay);
            }}
```

Java only allows comparison of strings using the == operator in very special circumstances. You can always use .equals to compare strings.

PRACTICE TASKS 5.3–5.5

5.3   The outline flowchart shown in Flowchart 5.5 matches one of the pseudocode fragments supplied in Options 1 or 2 for the exam grade calculation discussed in Skills Focus 5.2. Copy and complete this flowchart by adding pseudocode text for each of the symbols.

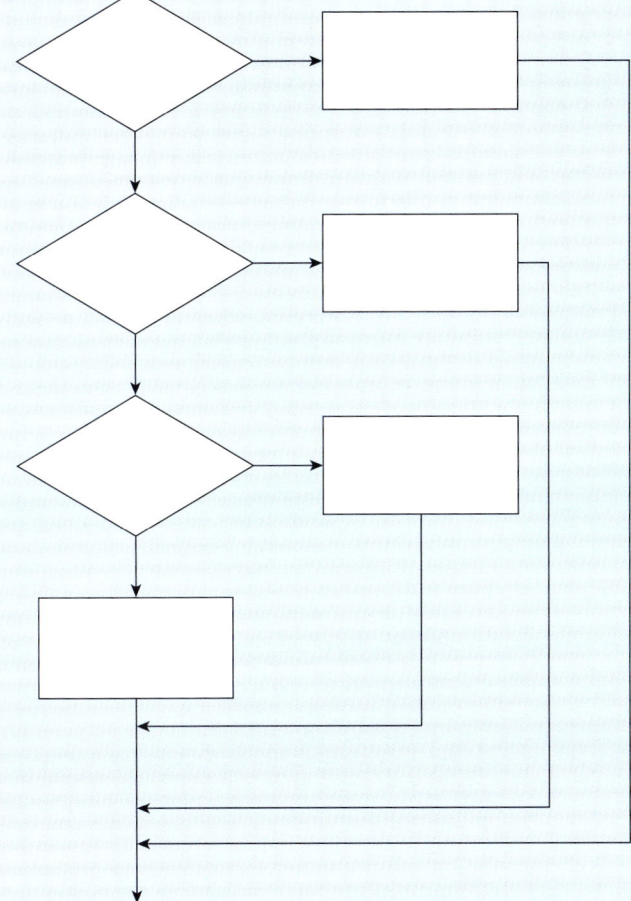

Flowchart 5.5: Flowchart for the exam grade program

5.4   Draw the equivalent flowchart fragment for one of the other solutions described in Options 1 or 2 in Skills Focus 5.2.

5.5   Create a Java program to incorporate your solution for either Practice Task 5.3 or Practice Task 5.4. The program should receive input of the exam mark and provide output of the grade.

5.5 CASE statements

A **CASE statement** is the alternative construct for a conditional statement. A CASE statement does not have the flexibility that you have seen is possible when an IF statement is used. However, when a CASE statement is an option, it is the preferable approach to take.

A CASE statement is an option if the actions to be taken in a program are dependent on the value of one variable.

Consider a program that functions as a simple calculator. The program would take as input two values and a further input to define the arithmetic operation to be performed.

To illustrate the programming construct without including too much code, just three options for the arithmetic operator will be used: we will define variables `number1`, `number2`, `answer` and `arithmeticOperator`. The variable `arithmeticOperator` can be given values 'A' for addition, 'M' for multiplication or 'S' for subtraction.

The following two pseudocode designs, Code snippet 5.1 and Code snippet 5.2, illustrate the simplicity of the CASE construct.

```
IF arithmeticOperator = 'A'
  THEN
    answer ← number1 + number2
  ELSE
    IF arithmeticOperator = 'M'
      THEN
        answer ← number1 * number2
    ENDIF
  ELSE
    IF arithmeticOperator = 'S'
      THEN
        answer ← number1 - number2
    ENDIF
ENDIF
```

Code snippet 5.1

```
CASE OF arithmeticOperator
    'A':   answer ← number1 + number2
    'M':   answer ← number1 * number2
    'S':   answer ← number1 - number2
ENDCASE
```

Code snippet 5.2

As you can see, Code snippet 5.1, which is using ELSE IF, is more cluttered than Code snippet 5.2, which shows the code for CASE. However, this is not the only benefit of using CASE. If you code with ELSE IF, the program has to read line by line until it finds a condition that is true. When a programming language supports CASE, the language incorporates software that allows a direct jump to the statement selected by the value of the variable.

In Java, the CASE statement is called a `switch` statement. The Java code corresponding to the pseudocode shown for the simple calculator program is:

```
switch (arithmeticOperator) {
    case 'A': answer ← number1 + number2;
    break;
    case 'M': answer ← number1 * number2;
    break;
    case 'S': answer ← number1 - number2;
}
```

Some points to note here are:

1 The code for what we can call the switch block is enclosed in braces { }.

2 The variable name must be inside parentheses ( ).

3 The break keyword causes an immediate exit from the switch code block to the statement immediately following the switch block. (The last option does not need break because exit follows automatically.)

Although there is a rigid structure for the Java CASE construct, there are some options. For example, it is allowed for each CASE example to lead to a number of actions. To achieve this, you can use braces to enclose a block of statements to be executed. For example, if you wished to include a different output for each option you might use:

```
case 'M': {
    answer ← number1 * number2;
    System.out.println("Multiplication gives " + answer);
    }
    break;
```

Following the last case in the list, it is possible to include a default action if none of the values defined in the cases has been supplied for the case variable. For example, the following could be included in the previous example:

```
default: System.out.println("unrecognised request");
```

The following code for the switch block will cause the program to output 'unrecognised request' if anything but A, M or S is input.

```
switch (arithmeticOperator) {
    case 'A': answer ← number1 + number2;
    break;
    case 'M': answer ← number1 * number2;
    break;
    case 'S': answer ← number1 - number2;
    break;
    default: System.out.println("unrecognised request");
}
```

Flowcharts are not uniquely used for creating algorithm designs; they are used in many different applications, so it is not surprising that there is no specific symbol for a CASE construct in a flowchart. Instead the diamond symbol has to be used.

One solution is to take the structure for the ELSE IF construct and adapt the text in the symbols. A slightly different solution is the one illustrated in Flowchart 5.6. The advantage of this style is that it emphasises that there is direct access to a specific case option.

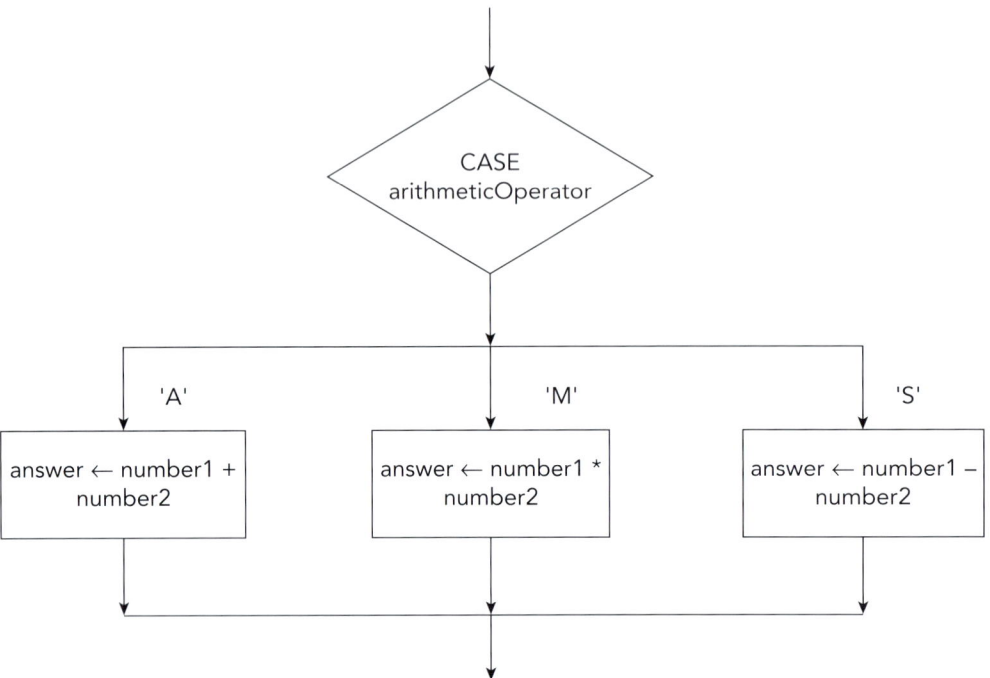

Flowchart 5.6: Flowchart example of a CASE construct

PRACTICE TASK 5.6

A ticket machine is to be installed at an airport. It will provide tickets for a railway that can take passengers to four stations. You have been asked to provide the Java code which will tell the user what they have to pay for a ticket.

The user has to identify the destination station for which the ticket is needed by entering 1, 2, 3 or 4. The ticket prices are:

- $10 for station 1

- $15 for station 2

- $20 for station 3

- $25 for station 4

You have decided to use a CASE statement in your Java code.

CHALLENGE TASKS 5.1–5.2

5.1 You are asked to provide the Java code for a program to provide a price for bus tickets. The requirement is related to the bus ticket price rules considered in Practice Task 3.5 and Challenge Task 3.1 of Chapter 3 and Practice Task 5.1 earlier in this chapter. However, the focus here is on assigning a price for one or possibly two tickets for one individual.

   1 The program should include a value of $50 stored for the price of a standard single ticket.

CONTINUED

2   The program should take as input an integer value for the age of the individual and a code for whether the ticket is either for a single journey or a single and return journey.

3   The program should then allocate a value for the category in accordance with the rules shown in Table 5.5.

Ticket type	Age	Category
Single	Age < 5	1
Single & Return	Age < 5	1
Single	Age >= 5 AND Age <= 10	2
Single & Return	Age >= 5 AND Age <= 10	3
Single	Age > 10 AND Age < 75	4
Single & Return	Age > 10 AND Age < 75	5
Single	Age >= 75	1
Single & Return	Age >= 75	1

Table 5.5: Category, age and ticket types for bus tickets

4   The program now needs a CASE construct with actions dependent on the value of the variable used to record the category. As children under 5 and adults aged 75 and over go free, Category 1 should be set at $0. Otherwise, the discount for a child is 50% of the adult price. The discount for a return trip is 20% of the single fare.

5   The program should output the price for the single ticket or for the single and return tickets. This can be done within the CASE construct.

5.2   A program is needed to calculate the amount of lawn seed needed to create a lawn.

The program must do the following:

- Take as input the shape of the lawn, which can be square, rectangular or circular.

- Take as input one or two dimensions (in metres) depending on the shape chosen.

- Take as input a decision as to whether to have a circular section left unseeded to allow roses to be grown in the middle.

- If the lawn is to have this unseeded circular area, calculate its area based on its diameter being one quarter of the width of the lawn.

- Take as input the amount of seed needed per square metre of lawn.

- Calculate the area to be sown.

- Calculate the amount of seed needed and output the amount.

Create a pseudocode or a flowchart design for the program. Then write the program in Java.

In this chapter, you have been introduced to the different possibilities for coding selection in a program. When you come to write a new program, you will have to decide on the appropriate choice every time the program requires a conditional statement. Figure 5.1 shows the sequence that your thinking should follow when making this choice.

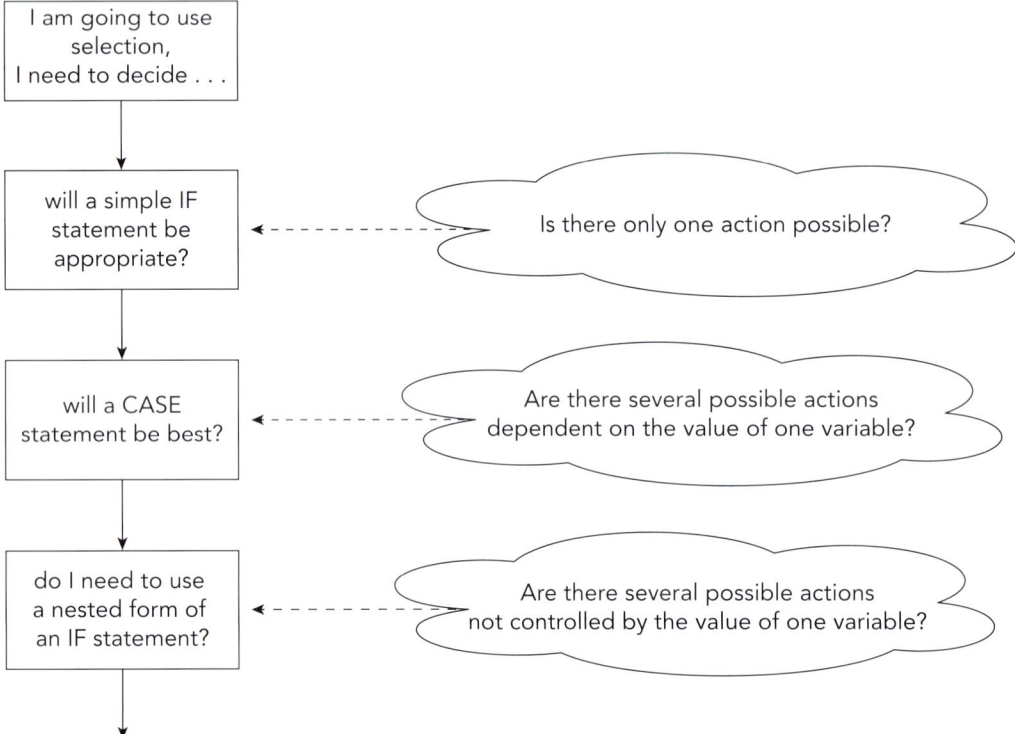

Figure 5.1: Choices to make when using selection

SUMMARY

Selection uses Boolean logic where true or false are the only values allowed.
A conditional statement defines a test of a logic proposition that is true or false.
A relational expression compares values using a relational operator.
Relational operators in Java are ==, !=, <, >, <= and >=.
A logic expression can include the Boolean operators NOT, AND and OR.
A logic expression often tests for a range of values.
An IF statement is the most common form of conditional statement.
An IF statement can be followed by an ELSE option or by an ELSE IF.
A CASE statement is an alternative form of conditional statement.
A CASE statement is the best solution if there are a number of options that are dependent on the value of one variable.

END-OF-CHAPTER QUESTIONS

1 Consider the following pseudocode:

```
IF value1 > value2
  THEN
    OUPUT "value1 is larger"
  ELSE
    IF value2 > value1
      OUTPUT "value2 is larger"
    ENDIF
```

....

....

....

Please supply the code for the three blank lines.

2 Provide the Java code to match the design in Question 1.

3 Consider a system that takes as input three different integer values. The system will output the highest value input. If a duplicate value is input, the system must output a message asking for a different value.

  a Design an algorithm for the system that uses a nested selection statement.

  b Implement and test your algorithm.

4 In a logic expression, the order of precedence for Boolean operators is:

  1 NOT has higher precedence than

  2 AND which has higher precedence than

  3 OR.

Consider the following logic expression where A, B, C and D are `Boolean` variables:

(NOT D) OR (A AND (NOT B)) OR (A AND C)

If B = true, which of A, C and D can have a value that guarantees a true value for the expression?

What is the other condition that leads to the expression being true?

> **Chapter 6**
Iteration

Introduction

Iteration is the final example of the three basic logic constructs introduced in Chapter 2. In that chapter, there were no details provided about how iteration would work. It was just indicated that it was often convenient to repeat a sequence of actions a number of times. This chapter will describe the different possible loop constructs that can control iteration in an algorithm. It will also discuss the factors that influence the choice of which one to use.

6.1 Iteration and loop constructs

Our lives are often subject to routine. Consider the following:

- turn off alarm
- get up and visit bathroom
- go to kitchen
- put kettle on and put bread into toaster
- consume toast and tea
- get dressed
- leave house in time to catch bus.

We could well repeat the same sequence of actions five times a week. This would be an example of iteration. There would normally not be any condition to check: we would just repeat the same actions for day 1, on to day 2, on to day 3, on to day 4 and finally on to day 5. However, for the weekend days or for holiday periods, these actions would not all be followed.

In programming, we rarely use the term iteration. Instead we talk about a loop, or a loop construct or looping. A **loop construct** can be categorised as being one of:

- A **count-controlled loop** where there is a defined number of iterations
- A **pre-condition loop** where the number of iterations is controlled by testing a condition at the start of the loop
- A **post-condition loop** where the number of iterations is controlled by testing a condition at the end of the loop.

These can be implemented as one of the following:

1   The '**For loop**', which is the only possibility for implementing a count-controlled loop. The number of repeats is defined in a heading for the loop with the understanding that no action inside the loop will change this. This would match a daily routine of the type described above. The heading could be 'For day 1 through to day 5':

   For day 1 through to day 5, repeat daily routine.

2   The '**While loop**', which is the only way of implementing a pre-condition loop. We could use this construct for our daily routine if the loop started with 'While it is a weekday':

   While it is a weekday, repeat daily routine.

3   The '**Repeat until loop**', which is one possibility for implementing a post-condition loop. This could be used for our daily routine if the loop started with 'Repeat' and ended with the condition 'Until it is the weekend':

   Repeat daily routine **until** it is the weekend.

KEY WORDS

iteration: a logic construct in which a sequence of actions in an algorithm is repeated several times.

loop construct: a coding for including iteration in an algorithm.

count-controlled loop: a type of iteration that will repeat a section of code a defined number of times.

pre-condition loop: a type of iteration where the decision to repeat a section of code depends on testing a condition at the start of the loop.

post-condition loop: a type of iteration where the decision to repeat a section of code depends on testing a condition at the end of the loop.

For loop: a coding for implementing a count-controlled loop.

While loop: a coding for implementing a pre-condition loop.

Repeat until loop: a coding for implementing a post-condition loop.

4 The 'Do while loop', which is the other alternative for a post-condition loop. This could be used for our daily routine if the loop started with 'Do' and ended with 'While it is a weekday':

Do daily routine **while** it is a weekday.

You can expect that any programming language will support a For loop construct and a While loop construct. However, the remaining two loops (Repeat until, and Do while) may or may not be supported. For the particular case of Java there is no Repeat until loop but there is a Do while loop as part of the language.

Note that the pseudocode that is used for presenting an algorithm design *does* include a Repeat until loop but *not* a Do while loop.

Table 6.1 gives examples of when each type of loop construct should be used.

KEY WORD

Do while loop: the other alternative for coding a post-condition loop.

TIP

It is vital that you understand that the condition in a Repeat until loop **stops the loop if true** whereas the condition in a Do while loop allows a further iteration if true and **stops the loop if false**.

Loop type	Description	When it should be used
For loop	Repeats a section of code a predetermined number of times.	The number of iterations is known or can be calculated. The programmer can set the code to loop the correct number of times.
While loop	Repeats a section of code while the control condition is true.	The number of iterations is not known and it may be possible that the code will never be required to run. The condition is checked before the code is executed. If the condition is false, the code in the loop will not be executed.
Repeat until loop	Repeats a section of code until the control condition is true.	The number of iterations is not known. However, the code in the loop must be run at least once because the condition is tested after the code has been executed.
Do while loop	Repeats a section of code until the control condition is false.	The number of iterations is not known. However, the code in the loop must be run at least once because the condition is tested after the code has been executed.

Table 6.1: The uses of loops

Let's suppose that you are writing a program and you decide that a loop is needed. Always begin by asking yourself several questions. Figure 6.1 shows how you could organise your thinking.

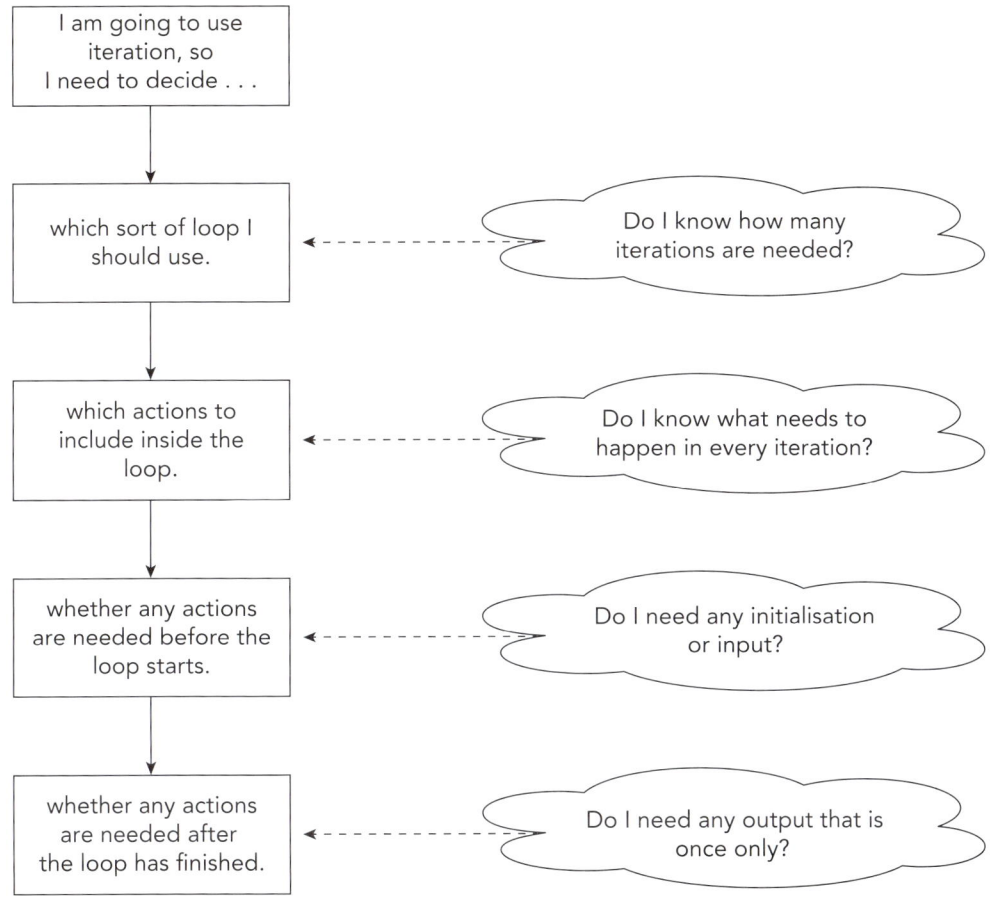

Figure 6.1: Questions to be asked when choosing a loop construct

We can now consider how the different loop constructs are coded and consider some examples of their use.

6.2 The For loop

The following is how a simple For loop can be included in a pseudocode design.

```
FOR i ← 1 TO 10
    <loop content>
NEXT i
```

Here we can see the main characteristics of a For loop:

1   The first line of code is the **loop header**.

2   The loop header identifies a **loop control variable**, which in this example is i.

3   The loop header also defines a range of values for the loop control variable.

4   The loop content follows the header; there is no limit on how many actions are defined here.

5   There is a feature that defines the end of the loop body; pseudocode uses NEXT to show the end of the loop body.

In this pseudocode, there will be successive iterations for values of i of 1, 2, 3, 4, 5, 6, 7, 8, 9, 10. The incrementation of the loop control variable uses a **step** of 1.

The Java code equivalent to the above pseudocode is shown here.

```
for (int i = 1; i <= 10; i++)
    {
    <loop content>
    }
```

Now the loop body is enclosed in a pair of braces. Therefore, a statement to indicate the end of the body is not needed. Table 6.2 contains explanations of the individual parts of the loop header.

> **KEY WORD**
>
> **step:** a value that defines how the loop control variable is incremented.

Java code	Explanation
`for`	The type of loop you are using. This must be in lower case in Java code.
`int i = 1;`	This statement is used just once as the loop begins. It is used to initialise the loop control variable `i`. It is usual to use single letter names for loop control variables. It is also good practice to confine their use to the one For loop. This means that they can be introduced in a declaration statement in a For loop header.
`i <= 10;`	This is the condition that is checked before each iteration to see if the loop can be entered again.
`i++;`	This is the definition of the step used. It states that the value of `i` will increase in steps of 1. `i++` is an increment operator. It is a concise representation of `i = i + 1`

Table 6.2: The loop header

Not all loops increase the value of i in steps of 1. Sometimes a loop is needed when the loop control variable is increased by other amounts. In pseudocode, the header must now contain a step definition following the definition of the range of values. For example, the header:

```
FOR i ← 1 TO 9 STEP 2
```

would cause i to take the successive values 1, 3, 5, 7, 9.

The equivalent Java code would be:

```
for (int i = 1; i <= 9; i = i + 2)
```

> **TIP**
>
> It is well worth carefully checking that the condition controlling the iteration is correctly formulated. If you make a mistake you will not get an error message, but the loop might never be entered – for example, if you wrote `i = 10` rather than `i <= 10`.

Two other points to note are:

1 The step can alternatively be defined as a negative value. For example, a pseudocode header could be:

```
FOR i ← 9 TO 1 STEP -2
```

Java also has an increment operator i-- to represent i = i - 1.

2 It is often stated that a For loop is used when the number of iterations is known. This is slightly misleading. It will only rarely be true that the number of iterations is fixed to be the same every time that the program is run. A better statement would be that 'once the loop has started, the number of iterations will be defined'. This is illustrated in Demo Task 6.1, where a variable is used in the part of the loop header that defines the range of values for the loop control variable.

DEMO TASKS 6.1–6.2

6.1 *You have been asked to produce a design and implementation for a program that will generate multiples of a specified number. For example, if the specified number were 7 and 10 multiples were requested, these would be:*

7, 14, 21, 28, 35, 42, 49, 56, 63 and 70.

Solution

Thinking about the requirement would lead to these decisions:

1 The program will be restricted to using integer values.

2 The multiples could be generated using a For loop.

3 The program would be rather useless if the multiples were not output.

4 The calculation of each multiple and its output must happen inside the loop.

5 The specified number and the number of multiples must be selected before the loop begins and nothing is needed after the exit from the loop.

You decide not to include declaration statements in the designs. Therefore, a suitable pseudocode design would be:

```
INPUT number
INPUT numberOfMultiples
FOR i ← 1 TO numberOfMultiples
    Multiple ← i * number
    OUTPUT Multiple
NEXT i
```

CONTINUED

The flowchart equivalent of this pseudocode is shown in Flowchart 6.1:

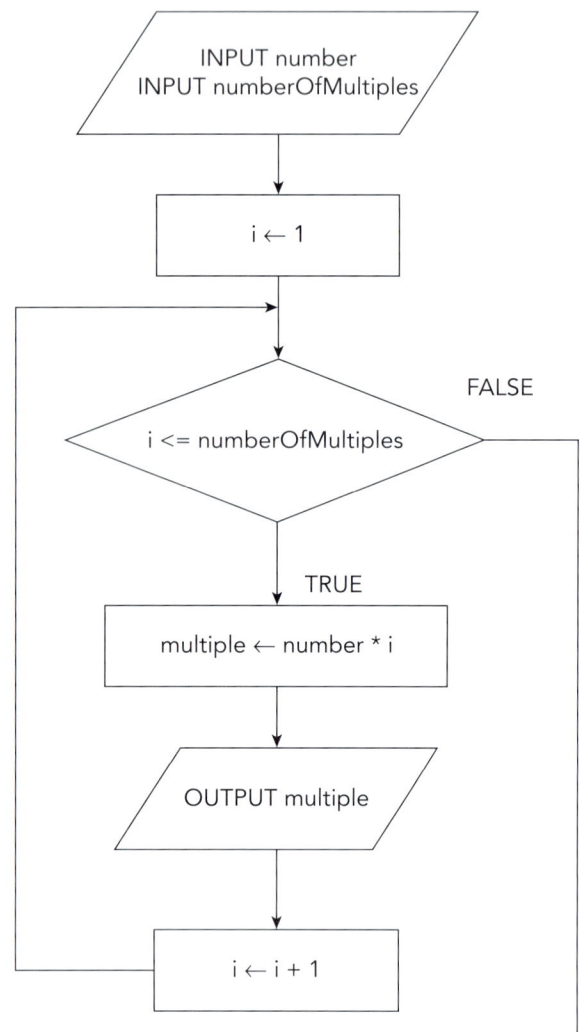

Flowchart 6.1: Flowchart for multiplication program

You may be surprised to see a decision diamond in the middle of this flowchart. The pseudocode version of the design appears to lack any decision. However, the programming language software that would implement a For loop does in fact test a condition before continuing with an iteration. This is made clear in the code used in Java for a For loop header. The loop control variable takes on a value beyond the defined range i <= 10 before the condition for continuation of the loop is tested. In this example, i is incremented to a value of 11 at the end of what becomes the final iteration.

CONTINUED

The Java code to match these designs for the multiples program is as follows. This is a complete program including declarations.

```java
import java.util.Scanner;
/*
the iteration example calculating multiples
 */
class Multiples {
public static void main(String[] args) {
    int multiple = 0;
    System.out.println("Please enter your number");
    Scanner numberInput = new Scanner(System.in);
    int number = numberInput.nextInt();
    System.out.println("Please enter number of
        multiples");
    Scanner multiplesInput = new Scanner(System.in);
    int numberOfMultiples = multiplesInput.nextInt();
    for (int i = 1; i <= numberOfMultiples; i++)
        {
        multiple = i * number;
        System.out.println(multiple);
        }
}}
```

6.2 *This time you have been asked to write a program that will total a series of values that will be input. You realise that a loop will be needed but you are not sure which type.*

Solution

You make an enquiry and you are told that the number of values will be known when the program is being used. Having given the problem some thought, you realise that the program will be best to use a For loop. The program will need to have these features:

- Input of the number of values before the loop is entered.

- Initialising of the sum to zero before the loop is entered.

- Input of a value inside the loop for each iteration.

- Incrementing of the sum inside the loop for each iteration.

- Output of the sum after the loop has completed.

You produce the following pseudocode design for the program (again you decide to ignore declaration statements).

```
total ← 0.0
INPUT valuesNumber
FOR i ← 1 TO valuesNumber
    INPUT value
    total ← total + value
NEXT i
OUTPUT total
```

CONTINUED

Flowchart 6.2 shows the flowchart version of the design:

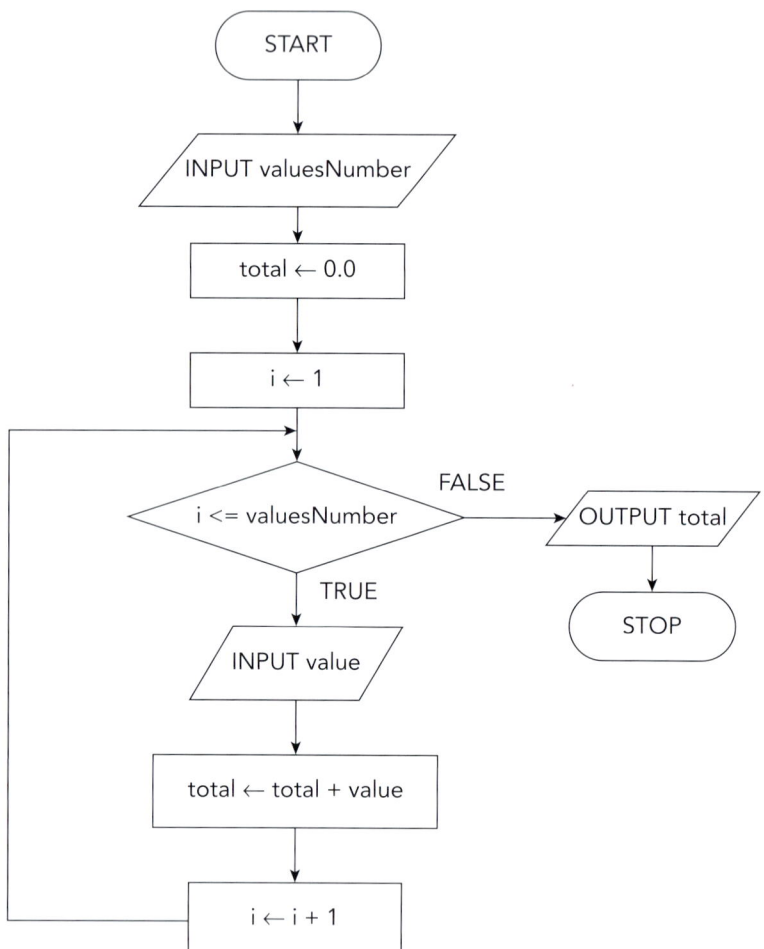

Flowchart 6.2: Flowchart design for the totalling values program

The following is an example of Java code suitable for this problem.

```
import java.util.Scanner;
/*
summing values
 */
class ValueSum {
public static void main(String[] args) {
   System.out.println("Please enter the number of
      values");
   Scanner numberInput = new Scanner(System.in);
   int valuesNumber = numberInput.nextInt();
   double total = 0.0;
   double value = 0.0;
```

CONTINUED

```java
for (int i = 1; i <= valuesNumber; i++)
    {
    System.out.println("Please enter a real
        number");
    Scanner realInput = new Scanner(System.in);
    value = realInput.nextFloat();
    total = total + value;
    }
System.out.println("The total for the values
    is " + total);
}}
```

PRACTICE TASKS 6.1–6.3

6.1 Implement the Java code shown for the multiples program in Demo Task 6.1. Run the program and check that the correct output is given for your chosen input values.

6.2 Implement the Java code shown for the summing values program in Demo Task 6.2. Run the program and check that the correct output is given for your chosen input values.

6.3 In Practice Task 5.2 in Chapter 5, you wrote a program to input an exam mark as a percentage and output pass or fail. Now modify this program so it will handle a number of marks and output pass or fail for each one. You will need to use a loop for this task. You can assume that the number of marks to be input is known.

CHALLENGE TASK 6.1

In Practice Tasks 5.4 and 5.5 in Chapter 5, you were asked to create a flowchart design and write a program that would take as input an exam mark and output a grade. Extend this flowchart design and then the program to fulfil the following requirements:

- The program will take input one-by-one of ten marks.

- The program will convert each mark to a grade and output the grade.

- The program will keep a running total of the numbers of each of the four grades (Distinction, Merit, Pass, Fail).

- When all ten inputs have been received, the program will output the total number of each grade.

6.3 The While loop

The While loop uses a construct with a header defining a condition to be tested.

The version you would have presented to you in pseudocode is:

```
WHILE <condition> DO
    <actions>
ENDWHILE
```

The Java version is similar.

```
while (<condition>) {
    <action>;
    <action>;
    ....
}
```

The logic of the While loop is that a condition is tested. If the condition is not true, the loop is ignored and the program starts running again, starting with the first statement following the end of the loop. If the condition is true, the statements inside the loop are executed once. The condition is then tested again to see if a further iteration is allowed.

There are two requirements for successful use of a While loop:

1   **To start:** There must be a value or set of values before the loop is encountered that ensure that the condition can be true and the loop can start. It would not make sense to define a loop that could never ever be entered.

2   **To end:** There must be some content in the loop that ensures that the loop will terminate at some stage. If this is not true, an unwelcome **infinite loop** will have been created.

Flowchart 6.3 shows a representation of a structure for a flowchart that can be considered as generic for either a For loop or a While loop.

> **KEY WORD**
>
> **infinite loop:** a loop that will iterate indefinitely. An infinite loop has no content that allows the loop to finish.

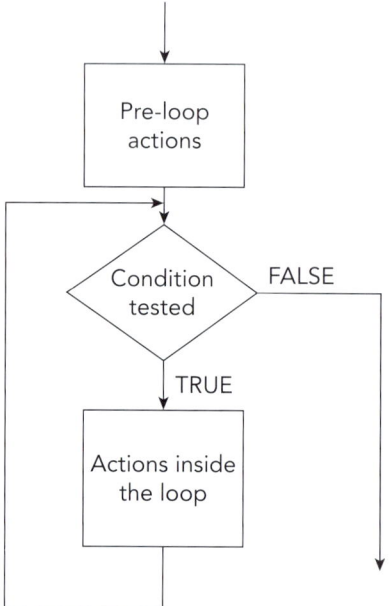

Flowchart 6.3: A generic flowchart for a For loop or a While loop

DEMO TASK 6.3

You have been asked to provide a program to total numbers which is slightly different from the one that was introduced in Demo Task 6.2. This program must be suitable for use when the number of values is not known when the program starts.

Solution

After thinking about the problem, you make these decisions:

• A While loop will be used.

• Since all of the numbers to be summed will be positive numbers, you can design the loop to continue until a negative number is input.

• The condition for the While loop will therefore be that a number input is greater than or equal to zero.

Your Java code solution for the main part of the program is shown here.

```
nextNumber = 0.0
total = 0.0
while (nextNumber >= 0.0) {
    total = total + nextNumber;
    System.out.println("Please enter a real number");
    System.out.println("To stop the program enter a
        negative number");
    Scanner realInput = new Scanner(System.in);
    nextNumber = realInput.nextFloat();
    }
System.out.println("The sum of the numbers is " +
    total);
```

You have made these decisions in this code:

• The negative number that stops the iteration must not be included in the sum.

• Therefore, you have made the input the last statement inside the loop.

• As a result, a positive number input is only added to the sum in the next iteration of the loop.

• This requires the sum statement to be the first one in the loop.

• The variable nextNumber is initialised to zero before the loop, which ensures that the loop is entered and that the value for total remains as zero until a value is input.

PRACTICE TASK 6.4

Create and run the Demo Task 6.3 Java program containing the While loop construct for totalling input values.

It is possible to use a While loop to achieve the same functionality as you would get if you used a For loop. In order to do this, the following are required:

- A loop counter variable that is incremented inside the loop.

- The While condition for the loop that references this loop counter variable.

- Initialisation of the loop counter variable before the loop begins.

- Sensible placing of the statement that increments the loop counter variable.

It is very easy to make mistakes in the logic. Practice Task 6.5 includes some examples of code that illustrate the problem.

PRACTICE TASK 6.5

The first row of Table 6.3 contains the code for a For loop and the output values produced when the code is run. The remaining four rows contain attempts to produce the same output using a While loop. For each one, write down what you would expect to see in the output column.

Note how the combination of the condition at the start of the loop and the positioning of the incrementing affect the output.

Pseudocode design	Output
```FOR i ← 1 TO 10 OUTPUT i NEXT i```	1 2 3 4 5 6 7 8 9 10
```i ← 0 WHILE i < 10 OUTPUT i i ← i + 1 ENDWHILE```	
```i ← 0 WHILE i < 10 i ← i + 1 OUTPUT i ENDWHILE```	
```i ← 0 WHILE i <= 10 OUTPUT i i ← i + 1 ENDWHILE```	
```i ← 0 WHILE i <= 10 i ← i + 1 OUTPUT i ENDWHILE```	

Table 6.3: Practice Task 6.5 loops

Can you create the pseudocode that produces an output of 1 through to 10 if `i` is initialised to 1 before the loop begins?

CHALLENGE TASK 6.2

This is a variation of a requirement for a program to calculate ticket prices for bus or rail travel. This scenario was featured in Challenge Task 5.1 in Chapter 5. This time you are asked to write a Java program that will allow details to be input for a number of people. When the input has finished the program must calculate and output a price for all of the tickets. The following is a partially completed pseudocode design for the program.

```
standardPrice ← 50
totalPriceToPay ← 0
INPUT age
WHILE age >= 0
    IF NOT (age < 5 OR age >= 75)
      THEN
        INPUT tickets
        IF tickets = 1 AND age >=5 AND age <= 10
          THEN
            category ← 1
        ENDIF
        IF tickets = 2 AND age >=5 AND age <= 10
          THEN
            category ← 2
        ENDIF
        IF tickets = 1 AND age >10
          THEN
            category ← 3
        ENDIF
        IF tickets = 2 AND age >10
          THEN
            category ← 4
        ENDIF
    ENDIF
    .........
    .........
    INPUT age
ENDWHILE
```

Following the setting of a value for category, the program can use a CASE construct to determine the price for the ticket or tickets (see Chapter 5). These rules apply:

- If tickets = 2 the second (return) ticket is discounted to 50% of the standard price. (This can be described as buy one get one half price).

- A ticket for a child is 25% of the standard price. The `totalPriceToPay` must then be incremented.

The program should be written so that the CASE construct is not used if age is under 5 or 75 or over.

You should note that the loop is terminated by the input of a negative value for age. You can use integer values for the variables.

6.4 The Do while and Repeat until loops

Once again there is a construct which includes a condition but now the condition is included in the statement at the end of the loop. The following is how a Repeat until loop would be presented to you in pseudocode.

```
REPEAT
     <actions>
UNTIL <condition>
```

The Java code for a Do while loop is:

```
do {
    <action>;
    <action>;
    . . .
    }
while (<condition>);
```

Flowchart 6.4 shows a representation of a structure for a flowchart that can be considered as generic for either a Do while loop or a Repeat until loop.

The following are some important points to remember:

- The actions within the loop are always performed at least once.

- The condition in a Do while loop is tested to see if the iteration is to continue; the actions continue while the condition is true.

- In contrast, the Repeat until loop ends if the condition is true.

- Java does not support a Repeat until loop construct.

The following is Java code for summing numbers using a Do while construct.

```
total = 0.0
nextNumber = 0.0
do {
    total = total + nextNumber;
    System.out.println("Please enter a real number");
    System.out.println("To stop the program enter a
        negative number");
    Scanner realInput = new Scanner(System.in);
    nextNumber = realInput.nextFloat();
    }
while (nextNumber >= 0.0);
System.out.println("The sum of the numbers is " + total);
```

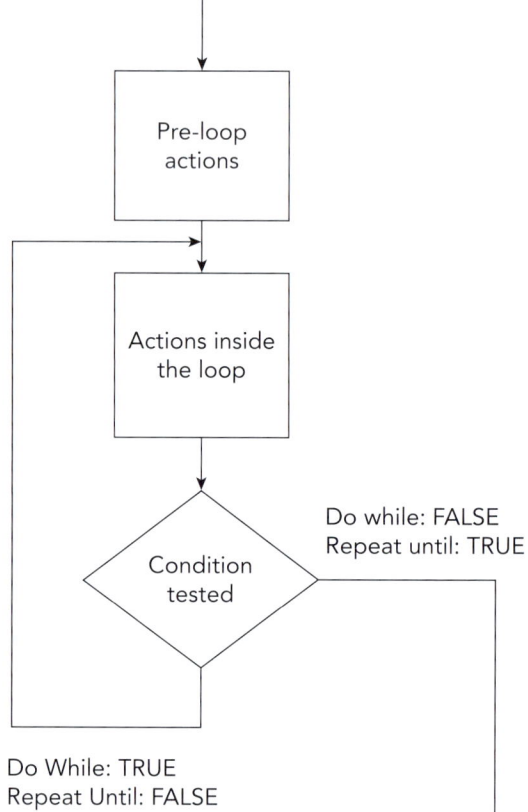

Do while: FALSE
Repeat until: TRUE

Do While: TRUE
Repeat Until: FALSE

Flowchart 6.4: Structure for a generic flowchart for either a Do while loop or a Repeat until loop

Note the following about the above Java code:

- Once again the negative number must not be included in the total.

- This can be achieved by having the input as the last statement in the loop body.

- The first time the loop is entered, two zero values are added.

6.5 Exiting a loop

The Java language does include a feature to exit a For loop before the defined number of iterations have been completed. However, it is often argued this is not best practice. At this stage of your learning, it is best if you only choose a For loop when you will definitely complete all iterations.

The examples included in Practice Task 6.5, earlier in this chapter, show that a conditional loop may be used with a number of iterations defined. However, you are more likely to use a conditional loop when there is an expectation that there will be an exit from the loop when a specified condition is met. It is worth considering different examples of this kind of use.

SKILLS FOCUS 6.1

OPTIONS FOR EXITING A LOOP

There are a number of reasons why a loop with the capacity to continue with further iterations will instead be instructed to finish looping. Some examples are considered here:

1   We have already shown the use of a negative number to stop further input. This is an example of a **rogue value**. It has to be a value that could not be a sensible input. It is the normal approach when values are being successively entered at the command prompt. The user of the program has no more values to enter. The loop has been written with a suitable condition for the rogue value to cause loop exit.

2   If data is being read from a file, there is no user in immediate control; instead, the computer accesses the file from within the program. It is possible for the file to have been created with a known number of lines, or with the last line of the file containing a rogue value (for example, −1) to signal the end of the file. However, a more reliable approach is to allow the program to check if any data remains unread from the file. If no data remains, the loop exits.

3   A program may be written to search for a particular value. There will be a need to exit the loop once the value has been found as there is no point in the program having to continue searching afterwards. This will be discussed in more detail in Chapter 8.

4   A variation on this would be when a user was playing a computer game. The game could involve successive input from the user. The loop would exit if the player managed to input the appropriate choice to win the game.

5   A program may be testing an assertion and it becomes clear that the assertion cannot be correct. There is no point in continuing. An example is considered in the following requirement.

KEY WORD

rogue value: a value entered by a user to stop the looping, for example, −1.

CONTINUED

Let's consider that a program is needed to check whether a number is a prime number. To understand the logic, you need to remember the definition of the modulus operator which was introduced in Chapter 3.

If x and y are two integer values, then the expression:

```
x modulus y
```

returns a value which is the remainder if x is divided by y.

For example, 35 modulus 20 returns a value of 15 (35 ÷ 20 = 1 remainder 15).

A prime number can be divided only by 1 or by itself. Therefore, the division by any other number will give a non-zero value for the modulus. This can be incorporated into an algorithm. The simplest design for such an algorithm uses a loop control variable that increases in steps of 1 from 2 up to one less than the number being tested. In Java, the loop will contain an assignment statement:

```
theModulus = numberToBeTested % i
```

where:

- `numberToBeTested` has been assigned a value before the loop starts using an input statement
- `%` is the Java version of the modulus operator
- `i` is the loop control variable.

The loop will be written so that if the value for `theModulus` is zero, the loop stops because `numberToBeTested` is not a prime number.

Questions

1 Write the Java program and run it to check it works properly.
2 In example 4 in the list of options for exiting a loop, the loop must stop when the player has entered a winning choice. How would you write code to achieve this?

SUMMARY

Iteration is the repetition of a sequence of actions.
Iteration is included in an algorithm by using a loop construct.
A For loop is a count-controlled loop construct.
A For loop uses a loop control variable for which the values are incremented for each iteration.
In a For loop, the number of iterations is defined once the loop has been entered.
While, Do while and Repeat until are three sorts of condition-controlled loop constructs.

CONTINUED

In a condition-controlled loop, the condition is tested in each iteration.
It is possible that a While loop is not entered at all.
If a Do while or Repeat until loop is used, the loop must be entered at least once.
When designing an algorithm using a loop construct, it is vital that you consider which actions come before the loop starts, which actions are inside the loop and which actions follow completion of the loop.

END-OF-CHAPTER QUESTIONS

1 a State the names of the three loop constructs that can be used in a Java program.

　b Give an example of a loop header for each of the three loop constructs.

　c Describe how a condition is used for each of the three loop constructs.

2 Describe three ways that a loop can be exited.

3 a Write the pseudocode for a test of whether an integer value is odd or even.

　b Write a pseudocode design for a program that inputs integer values and counts the number of odd values and the number of even values.

　c Explain how your pseudocode limits the number of input values.

4 Design and create a program that will take as input a series of positive numerical values greater than zero. The user will be repeatedly prompted to input numerical values until they enter the value zero to end the input stream. The program will then output:

　a the average of all the values input

　b an adjusted average; the program will ignore the largest value input and then calculate the average of the remaining values.

For example, if the input values are 20.5, 10, 35, 14.5 and 20, the value 35 is ignored. The average calculation is then (20.5 + 10 + 14.5 + 20) / 4 = 16.25.

> **TIP**
>
> It will be sensible to check for the maximum value as each value is input.

Chapter 7
Subroutines

IN THIS CHAPTER YOU WILL:

- understand the concept of a subroutine as a component of a program which can be accessed whenever it is needed

- learn how a subroutine is coded

- learn how a program accesses a subroutine

- understand the difference between a procedure and a function

- learn about the scope of variables

- learn about some pre-defined subroutines that are provided for random number generation, mathematics, string handling and file handling.

Introduction

A **subroutine** is a block of code that is written once and can then be accessed many times in a program. This chapter will show how subroutines are coded and accessed. It will introduce the terminology associated with the use of a subroutine. We will consider how the use of variables is dependent on where they are declared. The chapter concludes with some examples of the use of pre-defined subroutines.

7.1 Terminology

Chapter 9 looks at a modular approach to program design. In this approach, the design consists of a main program that can access a number of modules. At the final stage of the design process, there is a detailed design created for each module. When it comes to implementing such a design, program code for each module can be written in the form of a subroutine. This process is illustrated in Figure 7.1.

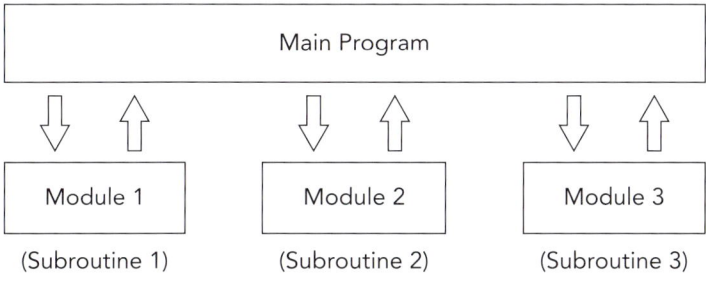

Figure 7.1: The concept of modular design

In this book we will use the term 'subroutine' as the generic name. We can then classify a subroutine as being either a **procedure** or a **function**. In Java and in object-oriented programming, a subroutine is called a **method**.

Before we look at the details of subroutine use it will be useful to consider the following overview of how they are used.

A subroutine, whether it is a function or a procedure, consists of a block of code that can be executed in a program whenever it is needed. However, the way that this block of code is used is different depending on whether it is being used as a procedure or as a function.

When a program is running and reaches a point where it needs to use a procedure the lines of code in the procedure are in effect inserted into the program code as extra statements. Once these have been executed the program code continues where it left off.

A program only uses a function during the execution of a single statement. The function is used as part of the expression in an assignment statement. It supplies a value to be used in the execution of the assignment statement. In providing the value for use in an assignment statement the function is said to **return** the value. The code for the function must include a statement that creates this return value and ensures that execution of the function code makes the value available for use by the program.

An important part of the use of a function or a procedure is the provision for a value already established in a program to be made available for the subroutine to use. A **parameter** is used for this. A parameter can also be used for the reverse process where a value established in a subroutine is made available for use by the program when this continues to be executed following use of the subroutine.

KEY WORD

parameter: a variable to be used in the subroutine. It is identified in the subroutine header.

7.2 Subroutine use

There are five main reasons for using subroutines:

1   A large program written as one piece of code will inevitably represent a complex problem. This applies when a program is initially written. It would also cause unnecessary difficulties for anyone having to correct errors in the code or introduce additional features later.

2   A subroutine can be developed, tested and maintained as an independent section of code.

3   Once a subroutine has been written, the code can be accessed in many different places in a program.

4   A subroutine can be used in different programs.

5   Libraries of thoroughly reliable subroutines are available so a programmer does not need to waste time writing a version every time if a suitable library routine is available.

It is important that you can distinguish between how a subroutine is defined and how a program can make use of a subroutine.

How a subroutine is defined

Before looking at how a subroutine could be created in Java we can consider the principles involved and the decisions that have to be made by reference to examples written using pseudocode. We will consider a subroutine that will check that the sum of three numbers is not greater than a maximum value allowed. The subroutine will ensure that a suitable message is output.

For a procedure definition the first line of code must be the header. A pseudocode example is:

```
PROCEDURE myProc(value1: INTEGER, value2: INTEGER, value3: INTEGER)
```

There are three elements here. The first has to identify that this is a procedure and not a function. The second is the name. A procedure is just like a variable in that it must have a name and it must be created before it can be used. The final element is a list of parameter names and their types enclosed in parentheses. The number of these can be zero or any higher number.

The definitions of the parameter names and data types in the procedure header act as variable declarations. The first line or lines that follow may contain additional declaration statements for variables. For this example we will just include one declaration:

```
DECLARE maximumNumber : INTEGER
```

The body of the procedure now includes whatever statements are needed. There are no restrictions on what can be used. For this example the following statements are included:

```
INPUT maximumNumber
total ← value1 + value2 + value3
IF total > maximumNumber
   THEN
     OUTPUT "There are too many"
   ELSE
     OUTPUT "numbers are within the defined limit"
ENDIF
```

The following should be noted about this code:

1   This example has an input statement inside the procedure body.

2   This is an option that is always available.

3   The alternative is to have the input statement in the program that is using this procedure.

4   Then the procedure will have to include a parameter that is to be used to receive this input value so that it can be used in the procedure.

Finally there must be a way to identify that there is no more code in the procedure body. Pseudocode uses the following statement:

```
ENDPROCEDURE
```

The above functionality could be achieved using a function rather than a procedure. The following pseudocode could be used for the function definition:

```
FUNCTION myProc(value1: INTEGER, value2: INTEGER,
  value3: INTEGER, maximumNumber: INTEGER) RETURNS STRING
     DECLARE decision: STRING
     total ← value1 + value2 + value3
     IF total > maximumNumber
       THEN
         decision ← "There are too many"
       ELSE
         decision ← "numbers are within the defined limit"
     ENDIF
     RETURN decision
ENDFUNCTION
```

The following should be noted about this code:

1   The function header (which has continued on to a second line) now defines a data type for the value that a function must return.

2   It has been decided to obtain the value for maximumValue by using a parameter rather than an INPUT inside the function.

3   The INPUT will have been included in the program code before this function is used.

4   The body code has been modified to use a string variable to hold the statement that in the procedure had been output.

5   A return statement is included to supply this statement to the program using the function. A function must have such a statement.

6   In the program using the function there must be an assignment statement using the function to provide the decision statement.

7   The program will then have an output statement for this decision.

How a program makes use of a subroutine

In the program that is using the subroutine, there is what is referred to as a **call** to the subroutine. If the subroutine has been defined with parameters the call of the subroutine must supply values to match these parameters. A value supplied as part of a subroutine call is referred to as an **argument**.

If a program is using a procedure, the call is a single action that requests that the code in the procedure body is run. An example of the pseudocode for calling the procedure that we have defined is:

```
CALL myProc(3,4,7)
```

If a program is using a function, the call to the function is always part, or perhaps the whole, of an expression on the right-hand side of an assignment statement. An example of the pseudocode for calling the function that we have defined is:

```
message ← myProc(3,4,7,100)
```

Defining a subroutine in a program design

Let's suppose that you are creating a design for a program and for a subroutine used by the program. In a pseudocode design, the coding for a subroutine definition must be created as a separate block of code. In a flowchart design for a program that contains a call to a subroutine, there must be a separate flowchart for the subroutine design. In the main program flowchart the following symbol, shown in Figure 7.2, is then used to represent this call:

Figure 7.2: This shape represents a subroutine call

The principles of the use of subroutines in a program can now be illustrated by examples of how Java handles them.

7.3 Java use of subroutines

Demo Tasks 7.1 and 7.2 show how you could write a subroutine and include it in a program.

DEMO TASKS 7.1–7.2

7.1 *You have been asked to provide code for the simple calculator program that was discussed in Section 5.5 of Chapter 5. This program performs arithmetic on two numbers*

Solution

You can see that a procedure can be written to contain the code for the CASE statement. Your thinking about what is needed is as follows:

- Three inputs are required from the user of the program.

- These are the two numbers and the arithmetic operator to be used.

- These could be input in the main program and supplied as arguments to the procedure call.

- This makes sense for the two numbers.

- For the operator you think that it will be a good idea to ask for input of 1, 2 or 3 to request addition, multiplication or subtraction.

- You decide that this can be done within the procedure.

- You realise that you can output the result of the calculation within the procedure.

The following is the code you develop. The colour coding is included as a basis for the explanations that follow.

```java
import java.util.Scanner;
/*
 using a procedure
 */
public class Calculatorproc{
static void calculator(int num1, int num2)
{
    int opcode = 0;
    int answer = 0;
    System.out.println("Please enter 1 for add,
     2 for multiply, 3 for subtract");
    Scanner myObj = new Scanner(System.in);
    opcode = myObj.nextInt();
    switch (opcode) {
       case 1: answer = num1 + num2;
    break;
```

CONTINUED

```
        case 2: answer = num1 * num2;
    break;
        case 3: answer = num1 - num2;
    }
    System.out.println("The answer is " + answer);
}


public static void main(String[] args) {
    int number1 = 0;
    int number2 = 0;
    System.out.println("Please enter first number");
    Scanner myObj1 = new Scanner(System.in);
    number1 = myObj1.nextInt();
    System.out.println("Please enter second number");
    Scanner myObj2 = new Scanner(System.in);
    number2 = myObj2.nextInt();
    calculator(number1, number2);
}}
```

The explanations for the code are given in Table 7.1.

Java code	Explanation

`import java.util.Scanner;`
`/*`
` using a procedure`
` */`
	The usual framework to allow the use of `Scanner` with your comment added.

`public class` `Calculatorproc{`	Because there is going to be a subroutine defined, this line needs `public class` rather than just class. This ensures that the subroutine can be used in the main program.
`static void calculator(int num1, int num2)`	This is the procedure header. It defines `calculator` as the name and `num1` and `num2` as parameters of type `int`. The use of `void` tells the compiler that this is a procedure – it does not return a value.

(continued)

TIP

When you write a program that includes the use of a subroutine you have written, you need to be careful to ensure that the braces are paired. It is easy to miss one off the end of the code.

CONTINUED

Java code	Explanation
| ```java
{
int opcode = 0;
int answer = 0;
System.out.println("Please
enter 1 for add, 2 for
multiply, 3 for subtract");
Scanner myObj = new
Scanner(System.in);
opcode = myObj.nextInt();

switch (opcode) {
    case 1: answer = num1 +
num2;
break;
    case 2: answer = num1 *
num2;
break;
    case 3: answer = num1 -
num2;
}
System.out.println("The
answer is " + answer);
}
``` | This is the code that makes up the body of the procedure.<br><br>Note that four variables are used. Two of these have been declared in the procedure header (`int num1` and `int num2`).<br><br>The other two are declared in the first two statements of the body code (`int opcode` and `int answer`).<br><br>The final statement provides output to the program user. |
| ```java
public static void
main(String[] args) {


}
``` | Unchanged framework holding the main program. |
| ```java
int number1 = 0;
int number2 = 0;
System.out.println("Please
enter first number");
Scanner myObj1 = new
Scanner(System.in);
number1 = myObj1.nextInt();
System.out.println("Please
enter second number");
Scanner myObj2 = new
Scanner(System.in);
number2 = myObj2.nextInt();
calculator(number1,
number2);
``` | This is the code for the main program. Within the program, you should note that the call to the procedure simply requires using its name with the arguments included in brackets:<br><br>`calculator(number1, number2);`<br><br>The arguments are the values stored for the variables `number1` and `number2`. These are integer values to match the integer parameters defined for the procedure (`num1` and `num2`). They have been supplied in the same order as in the parameter list in the procedure header. |

Table 7.1: Explanations for the code using a procedure
```

CONTINUED

As usual this code must be stored in a file where the name matches that for the class statement (in this case, Calculatorproc).

7.2 *You now wish to try out the alternative for the program you created in Demo Task 7.1. This is to write the program using a function rather than a procedure. The function will return a value to the main program.*

Solution

You decide that the output can be coded into the main program after the function has been used. Except for this difference, you intend to use the same approach with respect to how the subroutine receives data values.

The following code shows this alternative approach. The changed code is highlighted in brown.

```java
import java.util.Scanner;
/*
 using a function
 */

public class FuncCalc{
static int calculator(int num1, int num2)
{
    int opcode = 0;
    int answer = 0;
    System.out.println("Please enter 1 for add, 2 for
        multiply, 3 for subtract");
    Scanner myObj = new Scanner(System.in);
    opcode = myObj.nextInt();

    switch (opcode) {
        case 1: answer = num1 + num2;
    break;
        case 2: answer = num1 * num2;
    break;
        case 3: answer = num1 - num2;
    }
    return answer;
}

public static void main(String[] args) {
    int number1 = 0;
    int number2 = 0;
    int answer1 = 0;
    System.out.println("Please enter first number");
    Scanner myObj1 = new Scanner(System.in);
    number1 = myObj1.nextInt();
    System.out.println("Please enter second number");
    Scanner myObj2 = new Scanner(System.in);
```

CONTINUED

```
        number2 = myObj2.nextInt();
        answer1 = calculator(number1, number2);
        System.out.println("The answer is " + answer1);
    }}
```

Table 7.2 explains the reasons for the changes in the code.

Code change	Explanation
`static int calculator(int num1, int num2)`	This is the function header. As with a procedure, it contains the name and the declaration of two variables. The first difference is that there is no `void` included. This means that the subroutine will return a value, so therefore it is a function. Because it returns a value, it must have a defined type for this value. In this case the type is `int`.
`return answer;`	There has to be a statement inside the body of the function that defines the value to be returned. The subroutine has calculated a value for `answer`. The return statement ensures this is the value returned.
`int answer1 = 0;`	This declares the extra variable needed to receive the value returned by the function.
`answer1 = calculator(number1, number2);`	This is the assignment statement that uses the call to the calculator function. The values of the two variables `number1` and `number2` are supplied as arguments. The statement causes the value returned by the function to be stored in the variable `answer1`.
`System.out.println("The answer is " + answer1);`	The answer is now output in the main program.

Table 7.2: Explanation of the code changes needed when using a function

The approaches used in Demo Tasks 7.1 and 7.2 have achieved the same functionality. You may therefore ask why a programming language needs functions as well as procedures? The answers to this are:

- A procedure cannot return a value to the main program in the way that a function can.

- Therefore if the main program just needs to use a value calculated in the subroutine a function is chosen.

- If the only reason for using a subroutine is to output a result, then there is no point in using a function to return a value to the main program for the main program to output.

- A procedure can directly handle the output.

- If a procedure outputs a result, this informs the user of the program. However, the value is not available for subsequent use by the program.

TIP

It is important that you do not confuse returning a value with outputting a value. A value that is *output* is displayed for the user to see; a value *returned* is for the program to use.

7.4 The scope of variables

All programming languages have rules defining what is referred to as the **scope** of a variable. In the function code shown in Demo Task 7.2, there are two declarations in the body of the code. The variables declared are `opcode` and `answer`. These are two examples of a **local variable**. If a local variable has been created the following rules apply:

- The variable can only be used in the block of code containing the declaration statement for the variable.

- Any attempt to reference the variable elsewhere in the program (that is, in the main program or any other subroutines) will cause an error message.

However, this is not quite the full story. A different subroutine could be created. In this different subroutine, variables named `opcode` and `answer` could be declared. In this case, the uses of the two pairs of local variables would be entirely independent ones.

The concept of a local variable has been introduced in this chapter in the context of the coding for a subroutine. However, in Java you can create a local variable in any block of code, for example, in the body code for a loop construct. If there is a declaration inside the block, then the variable is local to that block. For example, Chapter 6 introduced a For loop giving the following code fragment as a framework:

```
for (int i = 1; i <= 10; i++)
    {
    <loop content>
    }
```

In this block, the loop variable `i` is local to the loop.

There is an alternative to using local variables. Most programming languages allow for a **global variable** to be declared. If a variable is declared as global, it can be referenced anywhere in the whole program (both in the main program and in any subroutines used). There are many sources that argue that it is bad programming practice to use global variables. For that reason, there are no examples in this book where they are used. It is just worth noting that Java does not define global variables as such. In Java, the same effect is achieved if the variables are declared as the first lines of code following the initial class definition in the framework we are using for running a program.

There is a related issue concerning the use of a subroutine with regard to the arguments supplied when the subroutine is called. If the argument is supplied as a variable in the call then the value currently stored for that variable is supplied as the argument. There are now two options which most programming languages support:

- The first is **call by value**. This is the method used in Java. When the code in the subroutine uses this value, any changes to the value stay in the subroutine. The variable in the calling program does not change. So, in the subroutines that have been coded earlier in this chapter with a header such as:

```
static void calculator(int num1, int num2)
```

the two parameters `num1` and `num2` become local variables.

KEY WORDS

scope: this defines the parts of a program where a variable will be recognised and can therefore be used.

local variable: a variable that is only recognised and usable in the block of code in which it is declared.

TIP

There can be many uses of the same name in one program providing that **a**, each one is a local variable with its own declaration statement, and **b**, you do not try to use that variable outside the subroutine you have declared it in.

KEY WORDS

global variable: a variable that is declared at the beginning of a program to make it accessible throughout the whole program code.

call by value: the action of providing a value to a subroutine that causes any changes to the value to be confined within the subroutine.

- The alternative is **call by reference**. In this case, each parameter in the subroutine is referencing the same memory location that the variable in the calling program is using. Any changes in the subroutine will change the value in the calling program. If call by reference is supported, there can be parameters used specifically for the subroutine to calculate values that the calling program can use following the use of the subroutine.

Java does not support call by reference. It does have a mechanism for something similar which does not involve using a variable value as an argument. However, this is beyond the scope of this book.

7.5 Using pre-defined subroutines

All programming languages provide what are referred to as library routines. These are subroutines that the creator of the language believes will be helpful to many programmers using the language. A **library routine** can therefore be used by any programmer. If a suitable library routine exists, you do not need to code one yourself.

If you intend to use a library routine, you need to understand the following:

- You will not have access to the code used in the body of the routine.

- Instead, you must read the documentation which is provided that tells you how the routine should be used and what precisely it will do.

- In particular, you need to find out whether or not the routine has parameters.

- If there are parameters, you need to know for each one what type of value is needed for the argument when the routine is called.

- Finally, you need to know whether the routine is built-in and available as part of the language or whether you must identify the library before you can use a routine in the library.

It is worth noting that libraries of routines are available that are not provided by the programming language installation. However, examples of these will not be considered in this book.

In Sections 7.6–7.9, we will look at examples of library routines provided by a Java installation that can be used for random number generation, maths calculations, string handling and file handling.

7.6 Random number generation

There are many possible applications for the use of random numbers. Examples include simulation programs, computer modelling and computer games. There is a vast literature on the subject of random number generation. You are well-advised to ignore this unless you have a keen interest in the topic.

We can now introduce a simple method (in other words, a subroutine) that Java provides. This method is associated with the Random class. It can only be used if this class is imported.

The code for this is:

```
import java.util.Random
```

The use of the random number-generating method (Random) involves a similar structure for the code created when Scanner is used for input. These lines of code are an example:

```
Random MyRand = new Random();
int randNum = MyRand.nextInt(50);
randNum += 1;
```

This code is explained in Table 7.3. There is a more detailed explanation of this type of coding in the Appendix.

Code	Explanation
Random	The name of the class that provides the method.
MyRand	An object created from the Random class.
int randNum	Declares a variable named randNum of type int.
= MyRand.nextInt(50)	Assigns a value to the variable randNum by using the nextInt method. The parameter 50 causes the random number to be one in the range 0 to 49 inclusive.
RandNum += 1	This ensures that the value is in the range 1 to 50 inclusive.

Table 7.3: Explanation of the random number generation code

DEMO TASK 7.3

You are interested in finding out just how random the numbers are when you use the Random method just described. You decide that you can do this by writing a program that repetitively uses the method and keeps a running count of the number of each value created. You expect that the numbers generated for each value should be approximately equal if the iteration is performed a large number of times. For the first attempt you decide to generate numbers in the range 1 to 6 and to try 100 iterations.

Solution

The program has a straightforward logic. Declarations are followed by a For loop. For each iteration of the loop a new number is generated. The total for that particular value is then incremented. The code is as follows:

```
import java.util.Random;
/*
the random number program
 */
class RandTest {
    public static void main(String[] args) {
```

CONTINUED

```
int sum1 = 0, sum2 = 0, sum3 = 0;
int sum4 = 0, sum5 = 0, sum6 = 0;
int randNum = 0;

for (int i = 1; i <= 100; i++)
{
Random MyRand = new Random();
randNum = MyRand.nextInt(6);
randNum += 1;

if (randNum == 1) sum1 = sum1 + 1;
if (randNum == 2) sum2 = sum2 + 1;
if (randNum == 3) sum3 = sum3 + 1;
if (randNum == 4) sum4 = sum4 + 1;
if (randNum == 5) sum5 = sum5 + 1;
if (randNum == 6) sum6 = sum6 + 1;
}
System.out.println("number of 1 values = " + sum1);
System.out.println("number of 2 values = " + sum2);
System.out.println("number of 3 values = " + sum3);
System.out.println("number of 4 values = " + sum4);
System.out.println("number of 5 values = " + sum5);
System.out.println("number of 6 values = " + sum6);

}}
```

The following output was produced when the program was run:

number of 1 values = 16

number of 2 values = 16

number of 3 values = 19

number of 4 values = 15

number of 5 values = 18

number of 6 values = 16

PRACTICE TASKS 7.1–7.3

7.1 Create and run the program developed in Demo Task 7.3. You can investigate the effect of using a larger number of iterations.

7.2 Create and run a program for a guessing game to be used by one person. The user is asked to input a number in the range 1 to 6. The program generates a random number in the same range. If the two numbers match the user receives a 'well done' message.

CONTINUED

7.3 Extend the functionality of the program you coded in Practice Task 7.2. The program needed now should allow the user to have ten attempts at guessing a number in the range 1 to 6. At the end of the ten attempts, the user gets a score out of ten showing the number of successful guesses.

QUICK QUESTION

What would you think could be classified as a good score when Practice Task 7.3 program was run?

CHALLENGE TASK 7.1

Create a program that could be used as a guessing game by two players. The requirements for the program are:

1   There will be ten rounds.

2   In each round the players are to enter a number in the range 1 to 100 inclusive.

3   Each number entered must be compared to a random number generated for that round.

4   The player who has entered a number closer to the random number gets 1 mark added to their score.

5   If the players' values are equally close to the random number you must repeat the process for that round.

6   After the tenth round, the winner is output.

You will need to think carefully about how you do the comparison with the random number.

7.7 Mathematics functions

Almost all programming languages are designed so that programs written in the language can be used in scientific or technical contexts. This means that there is a set of library routines devoted to applications in, for example, geometry, trigonometry or statistics where the use of mathematics is essential. Java has a Math library. Some examples of the methods in this library are given in Table 7.4 with some relevant comments about their use.

Method	How the method is used
sin() cos()	All of the trigonometric functions are available. The argument has to be supplied as a value in radians not degrees.
toRadians()	A useful function to do the conversion to radians for you. It is best to supply a value for the angle in the range 0 to 360.

(continued)

Method	How the method is used
sqrt()	Provides the square root of the argument value. Try to make this a positive value!
pow()	If, for example, you need to compute 3^4, some languages have an arithmetic operator to do this. Java has the pow function. The function has two parameters. The first argument supplies x in x^y and the second argument supplies y.
abs()	Supplies the absolute value; converts a negative argument to positive, leaves a positive argument unchanged.
round()	Converts the number to the nearest whole number.
hypot()	Calculates the hypotenuse of a right-angled triangle using the lengths of the other sides supplied as two arguments.
PI	This is not a function but it is provided as part of the Math library. You can use this instead of trying to remember the value of π.

Table 7.4: Subroutines in the Math library

You should note that most of these methods only work with data type double. Also the value for PI is stored as data type double.

The following program illustrates the use of a Math library method:

```
import static java.lang.Math.*;
/*
the demo of round
 */
class RoundTest {
    public static void main(String[] args) {
double number = 5.49;
double roundedNumber;
roundedNumber = round(number);
System.out.println(roundedNumber);
}}
```

Note here that the import line of code using * allows any of the methods in the Math library to be used in the program. The result output from running the above program is 5.0.

PRACTICE TASK 7.4

You have decorated a wall with a coat of paint. You intend to improve the decoration by painting on it ten squares in a different colour. Each square will be the same size, the same colour and fully filled in with a coat of paint. You have a tin of paint that will cover an area of 0.5 m². You are wondering how big the squares can be if you use this tin.

Write the pseudocode design for a calculation of the width of the squares that could be painted with this tin.

CHALLENGE TASKS 7.2–7.3

7.2 You have changed your mind. You have decided to paint circles rather than squares. Again, each circle will be the same size, the same colour and fully filled in with a coat of paint from this one tin.

Write the Java code for a calculation of the diameter of the circles that could be painted. The formula for the area of a circle is πr^2. Use the appropriate features from Table 7.4.

7.3 A program is needed to calculate the number of tiles needed to replace the roof on a building. The roof is symmetrical with two sloping halves as shown in Figure 7.3.

Figure 7.3 shows the dimensions for the building and the angle defining the slope of the roof.

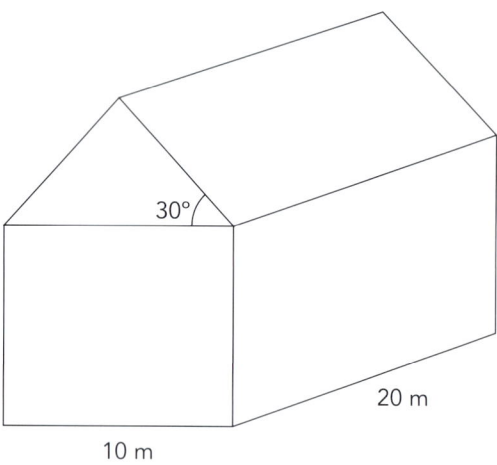

30°

20 m

10 m

Figure 7.3: The building that needs a new roof

The other information you need is:

- The tiles are rectangular with the vertical length 30 cm and the horizontal length 20 cm.

- The tiles are not to be cut; they are expected to overhang the wall at the bottom of the roof.

TIP

Carry out the calculation first for one of the sloping halves of the roof using a sensible choice of either the cosine function or the sine function.

7.8 String-handling methods (functions)

There is a fundamental difference between a value stored for a string compared with a value stored for, say, an integer. For a variable of an integer or a real type (sometimes referred to as 'primitive types'), the number of bits in memory needed to store a value is the same whatever the value. This is not the case for a string because strings can have different lengths (different numbers of characters). The Java solution for strings is to create them as objects.

There are three points you need to understand about the use of strings:

1   The type is `String` with an upper case S. (You do not need to worry about why!)

2   Individual characters in a string can be identified by an **index**. Table 7.5 illustrates this for a string "`123AbCdab`":

index	0	1	2	3	4	5	6	7	8
string	1	2	3	A	b	C	d	a	b

Table 7.5: A string index

> **KEY WORD**
>
> **index:** an integer value that identifies the position of a character in a string or an element in an array.

The important aspect of this is that the first character has index 0 not 1. In Chapter 8, you will see how an index is similarly used to identify an element in an array.

3   The fact that a string is an object affects the way that string-handling methods are used. The method is added after the variable name using a dot notation. As an example, the length of a string could be found using the following assignment statement:

        stringLength = theString.length();

where `theString` is the variable name and `length()` is the function.

Note that in this example, there are no arguments supplied but the paired parentheses must still be present to indicate that a method (in other words, a function) is being used.

A number of the methods available are described in Skills Focus 7.1.

SKILLS FOCUS 7.1

STRING-HANDLING METHODS AVAILABLE FOR USE IN A JAVA PROGRAM

A selection of the methods that Java provides are given in Table 7.6 with a brief definition for each. The examples that follow will give a more detailed explanation for each.

Method	Definition
`toLowerCase()` `toUpperCase()`	Convert the case of letters of the alphabet.
`charAt()`	Extracts a single character.
`length()`	Finds the length of the string.
`replace()`	Replaces characters.
`startsWith()`	Identifies the first character.
`substring()`	Extracts a substring (a section of a string).
`isEmpty()`	Checks if there are characters in the string.
`valueOf()`	Converts a value into a string.

Table 7.6: String handling methods and explanations

CONTINUED

Before looking at examples it can be noted that there is a concatenation method for joining two strings into one whole string. However, it is easier to use the + operator as illustrated by:

```
string3 = string1 + string2;
```

You can use the + operator if a String literal is too long to fit on one line. You split the string into two parts. The first line has the first part followed by + as shown in the following example:

```
string2 = string1 + "This is far too long a string" +
" to fit on one line so I have split it into parts";
```

Now we can look at the details relating to the use of these methods. For each method there is a fragment of Java code using the method. Each output is colour-coded in brown to distinguish it from the comments. Each example is based on the following declaration statement for the original string:

```
String mySTR = "123AbCdab";
```

Code using toLowerCase() and toUpperCase()

```
String mySTR1 = mySTR.toLowerCase();

System.out.println(mySTR1);

String mySTR2 = mySTR.toUpperCase();

System.out.println(mySTR2);
```

The output from these four lines of code is:
```
123abcdab
123ABCDAB
```
The important point to note is that any characters that are not alphabetic are ignored; they remain in the new string.

Code using charAt()

```
char myChar = mySTR.charAt(3);
System.out.println(myChar);
```
The output from these two lines of code is:
```
A
```
The important point to note is that the argument in the method is a value for an index so a value of 3 extracts the fourth character in the string (the first character has index 0). See Table 7.5.

Code using length()

```
int len = mySTR.length();
System.out.println(len);
```
The output from these two lines of code is:
```
9
```

CONTINUED

Code using replace()

```
        String mySTR3 = mySTR.replace("b", "zz");
        System.out.println(mySTR3);
        String mySTR4 = mySTR.replace("ab", "z");
        System.out.println(mySTR4);
```

The output from these four lines of code is:

```
        123AzzCdazz
        123AbCdz
```

The important point to note is that the arguments must be presented as strings even if there is only one character. The other point to note is that every instance of an argument in the string is replaced. In the first example, there are now two instances of zz.

Code using startsWith()

```
    String mySTR5 = mySTR.startsWith();
    System.out.println(mySTR5);
```

The output from these two lines of code is:

```
        1
```

Code using substring()

```
        String mySTR6 = mySTR.substring(4);
        System.out.println(mySTR6);
        String mySTR7 = mySTR.substring(3, 6);
        System.out.println(mySTR7);
```

The output from these four lines of code is:

```
        bCdab
        AbC
```

In the first example there is only one argument supplied. This is the index for the start of the substring. Every character up to the end of the original string is included in the substring. When a second argument is provided (mySTR7), this is the index following the one used to define the last character in the substring. This means that a second argument of 6 causes the sixth character in the original string to be included.

Code using isEmpty()

```
        boolean test = mySTR.isEmpty();
        System.out.println(test);
```

The output from these two lines of code is:

```
        false
```

A boolean variable test receives false as a value because the string is not empty. This method is useful for checking to see whether a user has responded to a request for a value to be entered.

Code using valueOf()

```
        int num = 35;
        String numstr = String.valueOf(num);
        System.out.println(numstr);
```

CONTINUED

```
    myChar = numstr.charAt(0);
    System.out.println(myChar);
The output from these lines of code is:
    35
    3
Here a variable storing an integer value is converted to a variable
storing a String value. The last two lines of the code are just there
to convince you that a String value has been stored.
```

Questions

1   In earlier versions of Java the isEmpty() method was unavailable.
    What alternative method could be used to check if a string were empty?

2   Can you write Java code to extract the last character in a string?

DEMO TASK 7.4

Suppose that you have some files containing Java programs. You have forgotten to ensure that the file extension is .java for some of the files. You are thinking of writing a program to change .txt to .java. You have decided to begin by writing a limited version of the program to test your understanding of the logic associated with string manipulation in Java.

Solution

The following is the code of this program:

```
import static java.lang.String.*;
import java.util.Scanner;
/*
changing file extension
 */
class fileExt {
public static void main(String[] args) {
    System.out.println("Please enter a filename with an extension");
    Scanner myObj = new Scanner(System.in);
    String fileName = myObj.nextLine();
    int fileLength = fileName.length();
    String ext = fileName.substring(fileLength - 4);
    if (ext.startsWith(".")) {
        String newName = fileName.substring(0, fileLength -4);
        String changedName = newName + ".java";
        System.out.println(changedName);
        }
    else System.out.println("no need to change name");
 }}
```

The string-manipulation methods have been highlighted in brown. Note that when substring is used with one argument, you automatically get all of the remaining characters up to the last position in the file. However, when you supply a second argument, this is exclusive. In this example, fileLength – 4 as the second argument ensures that characters from position 0 to position fileLength – 5 are included in the substring.

So if your file was called `123AbCdab.txt`, the line

```
String newName = fileName.substring(0, fileLength -4);
```

would only include `123AbCdab` in the string newName because in the filename 1 is at position 0 and b is at fileLength – 5.

You may find it helpful in this type of task to draw a table identifying the index positions as was shown in Table 7.5.

PRACTICE TASKS 7.5–7.6

7.5 In Chapter 4, you learned that if you are using `Scanner` for inputting data, you cannot use it directly to input a value to be stored in a variable of type `char`.

Write the Java code that uses the `charAt()` method to achieve input of a `char` value.

7.6 Suppose that data about individuals are stored in strings that have the following structure:

ID <space>first name<space>family name

The ID has four numeric digits. The first name and the family name each comprise a number of alphabetic characters. The number varies from person to person. For example:

1624 Aaron Lopez

Write a program that takes an input of an example of such a string, extracts the three components then outputs each one individually.

7.9 Using text files for input and output

When you are running a program, you will have data stored as values for variables and possibly large numbers of data values stored in arrays. All of this data will be lost when the program comes to completion.

For many programs, this is unsatisfactory. You will often wish to use a program time after time, knowing that data you used previously is still available for use each time. This can be achieved by storing data in files. The discussion here will concern the use of a text file that contains data stored as characters.

The following defines the terms and concepts that are associated with the use of files in Java:

1 Your system will have a filestore (a place that stores files that can be accessed by the program). In the filestore, each file has a name. An example name for a text file to be used by a program would be ProgFile.txt.

2 Your text files need to be in the folder or directory that is used to store your Java program files.

3   If you write a program that needs to access a text file, the program has to open the file.

4   When the program opens a file, it creates a name to be used in the program (instead of using the name used in the filestore). This name that the program creates is then used any time the program accesses or refers to the file in your filestore.

5   The action of opening a file also specifies if the file is open for reading from or open for writing to.

6   If the program opens a file for reading from, then the file must already exist in your filestore.

7   If the program opens a file for writing to, this file will be created in the filestore if it does not already exist.

8   When the program no longer needs to use a file, the program must close the file.

When writing programs in Java that use one or more files, you have many options. For text files, you need to choose the features that are specifically for reading or writing character data. You also have to consider exception handling. This needs a little explanation.

7.10 Exception handling

In Chapter 11, we will look at the problems that can occur when you write a program and attempt to run a program. The focus there is on errors in your code. These can be syntax errors that the compiler picks up. Alternatively, they can be runtime errors that happen because the program logic cannot handle the values you have chosen to use.

When you are using a program that uses a file, there is the possibility of a different kind of problem at the moment you open a file. The file may not exist or may not be in the folder or directory you are using, or there is a possibility that you are not allowed to create a file in the folder or directory. This problem is not referred to as an error but instead as an exception.

If you try to run a Java program that uses a file, but you have not included any reference to exception handling, you will receive an error message. This message will tell you that the program will not be run without some detail being included that relates to handling exceptions. The range of options for this are beyond the scope of this book. Just the simplest possible approach is considered here.

7.11 The Java framework for file handling

The following framework can be used when a program needs to read from a text file and write to a text file.

```java
import java.io.FileReader;
import java.io.FileWriter;
import java.io.IOException;
```

KEY WORDS

open: an action in a program that allows access to a file in your filestore.

close: an action that releases the file access when the program no longer needs to use it.

exception handling: the inclusion of some coding that defines what should happen if the program cannot do what has been requested, such as use a specific file that cannot be found.

```
import java.util.Scanner;
public class ** {
public static void main(String args[]) throws
    IOException   {
**
}}
```

The parts highlighted in brown are new features. FileReader is specifically used for reading character data; FileWriter is used for writing character data.

The import of IOException plus the addition of throws IOException to the use of main is the simplest coding to allow a program to run with the use of files.

The use of Scanner is not essential. Java provides many other options, but it is a facility that you are already using regularly.

The following example shows how a line of text can be read from a file.

```
import java.io.FileReader;
import java.io.FileWriter;
import java.io.IOException;
import java.util.Scanner;
public class Fileread {
public static void main(String args[]) throws
IOException   {
    String strInput;
    FileReader finputObj = new FileReader("ProgFile.txt");
    Scanner InObj = new Scanner(finputObj);
    strInput = InObj.nextLine();
    System.out.println(strInput);
    finputObj.close();
}}
```

The following example shows how a string can be output to a file.

```
import java.io.FileReader;
import java.io.FileWriter;
import java.io.IOException;
import java.util.Scanner;
public class Filewritetest {
public static void main(String args[]) throws
IOException   {
    String OutString = "Trial string";
    FileWriter foutputObj = new FileWriter("ProgWriteFile.
    txt");
    for (int i = 0; i < OutString.length(); i++)
        foutputObj.write(OutString.charAt(i));
    foutputObj.close();
}}
```

PRACTICE TASK 7.7

Write a program that reads a line of data from a file, then writes the data back to a different file. You can use the coding provided in the examples that you have just read in Section 7.11 as the basis for your program.

SUMMARY

A subroutine is a component of a program which can be accessed whenever it is needed. Java uses the name method rather than subroutine.

A subroutine is said to be called when a program uses it.

The code for a subroutine contains the subroutine's name plus parameters if these are needed.

Subroutines include procedures and functions.

- A procedure is called by having a program statement containing just its name. A procedure does not return a value.

- A function is called in an expression used in an assignment statement and must be defined with a data type. It returns a value of the same data type as the function to the calling program.

The scope of a variable defines where in a program it can be used. This includes local variables and global variables.

Java provides pre-defined subroutines for creating random numbers and for carrying out mathematical calculations.

Many methods are available for extracting characters from strings and for changing the characters in a string.

Java provides methods for reading from a text file and for writing to a text file.

Java will not allow use of a text file inside a program unless exception handling has been specified in the coding.

END-OF-CHAPTER QUESTIONS

1 a Explain the difference between an argument and a parameter. You may answer this by providing example code.

b Explain three differences between a function and a procedure.

2 a Identify three methods available in Java for handling strings.

b Provide an example of Java code for each method.

c Explain how each method works, including whether or not arguments are required.

3 A method is needed that can be called from the main program. The method will convert a time measured in seconds to show the same time in minutes and seconds. The program will pass the method a single integer value that represents the number of seconds. The method must calculate the appropriate minutes and seconds. For example, if passed 190 seconds, the method would convert this to 3 minutes 10 seconds.

You can provide the solution as a pseudocode design or as Java code. The solution can use either a function or a procedure.

> Chapter 8

Arrays

IN THIS CHAPTER YOU WILL:

- understand the concept of a one-dimensional (1D) and of a two-dimensional (2D) array

- learn the terminology associated with the use of an array

- learn how to declare an array

- learn how to instantiate and initialise an array

- understand the typical uses of an array

- write code using arrays.

Introduction

An array is an example of a data structure. This chapter introduces the difference between the data structure for a one-dimensional array and a two-dimensional array. The terminology associated with the use of an array is described. We will learn how the use of an array simplifies coding, particularly when a program includes iteration.

8.1 Terminology and concepts

An array is described as a **data structure** because it stores more than one value. There is a single name to identify the array. Each data value is stored in an **element** of the array. Each element is identified by an **index**. All of the elements in the array must have the same data type. An element of an array can be used in exactly the same way as a variable can be used in a program. Table 8.1 shows an example of the storage of data in an array.

Array elements in an array named `weekName` **of type** `String`	Value stored
weekName(0)	"Monday"
weekName(1)	"Tuesday"
weekName(2)	"Wednesday"
weekName(3)	"Thursday"
weekName(4)	"Friday"
weekName(5)	"Saturday"
weekName(6)	"Sunday"

Table 8.1: Data values stored in the elements of an array

Note that the index is an integer value and the first element has index 0. You can also see one immediate benefit of using an array. Here there is just the one name (weekName). Using variables would require seven names: weekName1, weekName2 and so on.

8.2 Declaring, instantiating and initialising

If you are presented with a pseudocode design for a program using an array, the declaration statement is reasonably straightforward. However, the pseudocode syntax is designed to be of general use whatever the programming language. It does not map on to the Java syntax at all well. The generic version can be represented as:

```
DECLARE <identifier> : ARRAY[<lower>:<upper>] OF
<data type>
```

Here <lower> defines what is referred to as the lower bound of the array and <upper> the upper bound. These are the lowest and highest index values defined for the array.

The declaration in pseudocode for the weekName array illustrated in Table 8.1 would be:

```
DECLARE weekName : ARRAY[0:6] OF STRING
```

In Java, the declaration of an array is different to the declaration of a variable. This requires a two-part process. The construct is similar to the one you are using with Scanner, and was used in Chapter 7 when using Random.

Again, using the weekName array as an example, the first part is the declaration:

```
String[] weekName;
```

This works as a sort of declaration of intent to use an array. The second part:

```
weekName = new String[7];
```

is said to **instantiate** the array. It allocates the memory space so the array can be used. You should note that the value given in square brackets is the **length** of the array – in this case, 7 elements. The index values range from 0 to the length minus 1 – in this case, the index values would be 0 to 6. (The 'minus 1' is because the index starts at 0. If we were to count the length of the array, we would start at 1 for the first element.)

It is possible to combine the two statements into one:

```
String[] weekName = new String[7];
```

The corresponding declaration for an array containing integer values would be:

```
int[] dayNumber = new int[7];
```

The above is similar to the approach used by many programming languages in that the length of the array is explicitly defined. However, Java has an alternative approach, illustrated by the following example:

```
int[] dayNumber = new int[] {1, 2, 3, 4, 5, 6, 7};
```

This statement serves to declare and instantiate the array and to **initialise** the values stored using the following three parts:

* declared: `int[] dayNumber`

* instantiated: `dayNumber = new int[]`

* initialised: `{1, 2, 3, 4, 5, 6, 7};`

This is a useful concise construct for arrays with a few elements. You can see that the array contains seven elements, but this is not explicitly defined. When this type of construct is used in a program, there needs to be a way that the length can be found. Java achieves this by storing internally a variable `length` when an array is created. The value stored in this variable can be found in the program by using a statement such as:

```
int dayNumberLength = dayNumber.length;
```

Here the variable dayNumberLength has now been declared and initialised to contain the length of the array dayNumber.

KEY WORDS

instantiate: the action that causes memory space to be allocated for an array.

length: the number of elements in an array.

initialise: an action that provides a value for a variable before the variable is first used in a program.

TIP

It is vitally important that you do not confuse a value supplied for the length of an array in a declaration with a value used as an index when an array element is accessed in a program.

8.3 Iteration using arrays

There are two fundamental uses of iteration for dealing with an array. The first is for populating an array with values. Section 8.2 showed how values could be provided in the declaration statement. However, that approach can only be used in these circumstances:

- The programmer needs to store a specific set of values.

- The values are chosen to suit the particular application.

- The values need to be available immediately at the start of the program.

For example, a programmer might be creating a computer game where the target numbers are stored in an array.

More often, a program needs to store values that are provided or are generated during the running of the program. The following code fragment shows how a user could input 20 exam marks into an array. The 20 marks occupy the elements with index values 0 to 19.

```
int[] examMark = new int[20];
for (int i = 0; i <= 19; i++)
    {
    System.out.println("Please enter the next mark");
    Scanner myObj = new Scanner(System.in);
    examMark[i] = myObj.nextInt();
    }
```

Table 8.2 contains explanations of the syntax in this code fragment.

Code	Explanation
`int[]`	Defines the declaration to be for an array storing integer values.
`examMark`	The name of the array
`new`	Always needed for the creation of an object, which in this case happens to be an array.
`int[20]`	Confirms the integer type and defines the length of the array.
`for (int i = 0; i <= 19; i++)`	A For loop header that increments i from 0 to 19 to match the values for the index in the array.
`System.out.println("Please enter the next mark"); Scanner myObj = new Scanner(System.in);`	The usual prompting for input and use of Scanner.
`examMark[i] = myObj. nextInt();`	An array element being used in the same way as a variable would be.

Table 8.2: Explanations of the code fragment syntax

The other fundamental use of iteration is for outputting values that are stored in an array. As an example, the exam marks could have been converted to a percentage mark, which was stored in an array `examPercent`. The following code might be used for the iteration that outputted these percentage marks:

```
for (int i = 0; i <= 19; i++)
    {
    j = i + 1
        System.out.println("The percentage for candidate
            " + j + " is " + examPercent[i]);
    }
```

This code assumes that the marks were originally input in a defined order. The user of the program would know the identity of candidate 1, candidate 2 and so on.

In Chapter 6, we saw that an iteration construct can allow a program to enter a succession of values and use them in calculations. The program for Practice Task 6.2 in Chapter 6 is an example. A feature of that type of program is that each input overwrites the value that was input in the previous iteration. When all the input values need to be stored, an array provides the ideal solution.

DEMO TASK 8.1

You have been asked to develop a program to collect statistics calculated from a number of values that will be input to the program. You decide to design and then implement a prototype program to which you can later add extra features.

The prototype program will have these features:

- *The first input will provide the number of values that are going to be input in the iteration.*

- *This allows a For loop to be used.*

- *The program will handle integer values.*

- *The program will store input values in an array.*

- *The input values will be summed.*

- *The input values will be output, together with their sum.*

Solution

The following is a possible pseudocode design for the prototype program.

```
DECLARE howManyInput, sum : INTEGER
sum ← 0
INPUT howManyInput
DECLARE numIn : ARRAY[0: howManyInput - 1] OF INTEGER
FOR i ← 0 TO howManyInput -1
    INPUT numIn[i]
    sum ← sum + numIn[i]
NEXT i

FOR i ← 0 TO howManyInput -1
    OUTPUT numIn[i]
NEXT i
OUTPUT "Sum of above numbers is ", sum
```

CONTINUED

You have included a separate loop for outputting the numbers to allow maximum flexibility for extended versions of the program.

A flowchart does not need any special features when an array is used. Flowchart 8.1 shows a flowchart for the first part of the program.

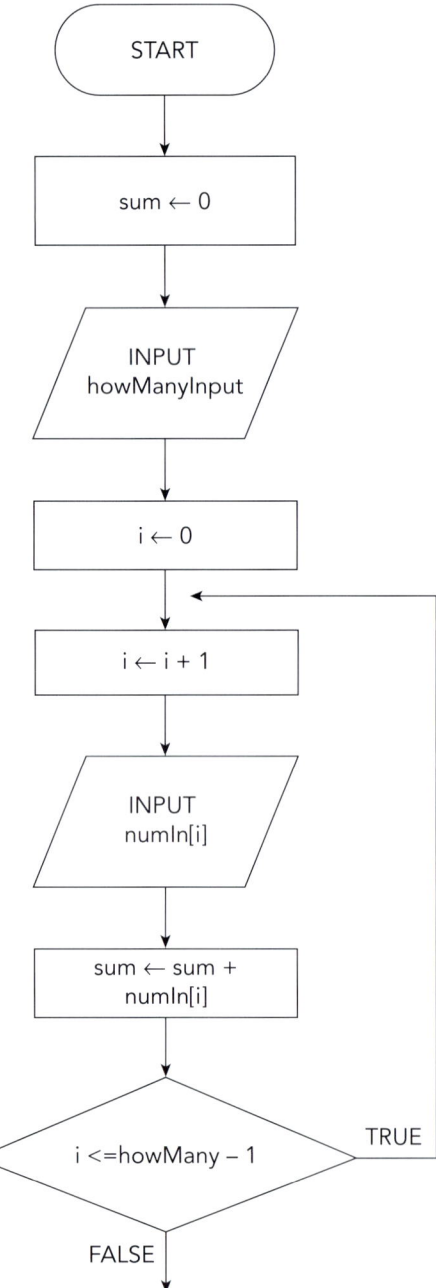

Flowchart 8.1: A flowchart for the first part of the program

CONTINUED

Table 8.3 shows the Java code for the program with some comments.

Java code	Comments
| ```java
import java.util.Scanner;
/*
summing using an array
*/
class arraySum {
public static void main(String[] args) {
``` | The usual header code. |
| ```java
int sum = 0;
System.out.println("How many numbers?");
Scanner myObj = new Scanner(System.in);
int howManyInput = myObj.nextInt();
``` | Initialisation then input. |
| ```java
int[] numIn = new int[howManyInput];
``` | The statement that instantiates the array numIn. Note that at the time the program is presented to the compiler, the length of the array is not defined. However, this statement now defines the array length ready for when the array is first used. |
| ```java
for (int i = 0; i <= howManyInput - 1; i++)
    {
    System.out.println("Number please?");
    Scanner myObj1 = new Scanner(System.in);
    numIn[i] = myObj1.nextInt();
    sum = sum + numIn[i];
    }
``` | A loop structure that inputs one value in each iteration and increments the sum.

The important point to note is the choice of the range of values for the loop variable. This is dependent on the fact that the index values for the array start at 0. |
| ```java
for (int i = 0; i <= howManyInput - 1; i++)
    {
    System.out.println(numIn[i]);
    }
System.out.println("Sum of above numbers is " + sum);
}}
``` | A second loop to output the individual values. The final sum is output after the loop. |

Table 8.3: Java code for the statistics program
```

PRACTICE TASK 8.1

Use the code from Demo Task 8.1 to implement and then run the program to check it works properly.

QUICK QUESTION

You are thinking there might be a need for a program to use a conditional loop to enter numbers into an array. This would be because the number of input values was not defined. Can you think of any problem that might occur with such a program?

PRACTICE TASK 8.2

A games program is to be based on the use of an array of values that have been generated using a random number generator. At this stage, you need to create the values for all of the elements.

Provide the program code for creating and storing the values in the array. You can create an array with ten elements holding integer values. The random number generation is described in Section 7.6 of Chapter 7. Generate random values in the range 1 to 20. Complete this preliminary program by including output of the values stored to check that they fit the requirement.

8.4 Using arrays in standard methods of solution

A **standard method of solution** is one that can be applied in many different applications. The following is a representative list of those commonly used:

- searching – where the algorithm is checking whether or not a particular value is present in a collection of values

- sorting – where the algorithm is rearranging a collection of values into a chosen order

- totalling – where the algorithm is taking each value in turn and adding it to a running total

- counting – where the algorithm is incrementing a count each time a further value is found in a collection of values

- finding maximum or minimum values – where the algorithm is checking if a value is, respectively, the highest or lowest so far each time a further value is found in a collection of values

- finding an average value – where the algorithm is totalling values before calculating the average.

KEY WORD

standard method of solution: a generic technique that can be applied in many applications.

All of these need to use iteration. They all benefit from the use of an array to ensure efficient coding. You have already met examples of totalling, counting and finding maximum, minimum and average values. We now need to consider searching and sorting.

Linear search

When an array has been created and populated with values, there is often the need for a program to check if a particular value has been stored in the array. The process of checking for a value is described as searching. The **linear search** is the simplest method for searching. It can be used when data values have not been sorted into order.

The program will require an iteration. At each iteration, the **search value** is compared with the value stored for an element.

There are two possibilities for the use of the search. In the first one, the aim is to count how many times the value is stored in the array. In the second one, the array is known to have no duplicate values. The search will simply confirm whether or not one of the elements matches the search value.

KEY WORDS

linear search: a technique where values in an array are successively compared to a search value.

search value: a value that you are looking for in the stored array elements.

TIP

It is possible to use a linear search to find a value stored in a text file, but the coding is not so simple.

DEMO TASK 8.2

Consider that you need a program to search for a value in array. If it is known that all of the values in the array are different, then once the value has been found, the search needs to stop. The search should also finish if all the values in the array have been checked.

Solution

The following is a pseudocode design for such a linear search program. The code is for an array containing 10 elements.

```
INPUT nameSought
found ← FALSE
index ← 0
WHILE found = FALSE AND index <= 9
    IF nameList[index] = nameSought
      THEN
         Found ← TRUE
         OUTPUT "Item was found"
      ELSE
         index ← index + 1
    ENDIF
ENDWHILE
IF found = FALSE
  THEN
     OUTPUT "Item was not found"
ENDIF
```

CONTINUED

Flowchart 8.2 shows the equivalent flowchart design:

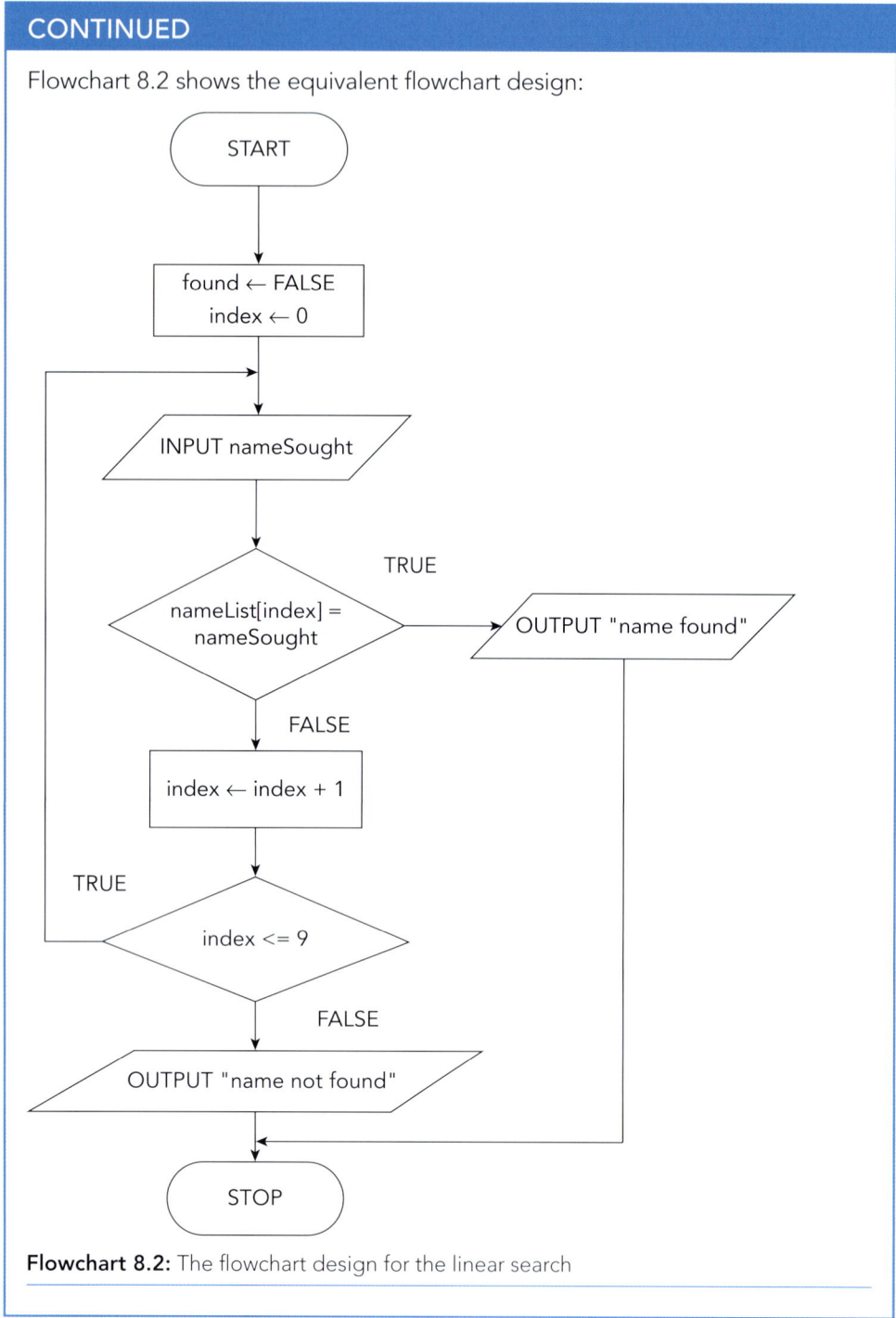

Flowchart 8.2: The flowchart design for the linear search

PRACTICE TASK 8.3

Write a Java program to implement a linear search based on the designs shown in Demo Task 8.2. You will need to input some values and store them in the array. You can limit the number of values. You may choose to use `int` values rather than `String` values. Run the program with a value that should be found then run it, once again with a value that will not be found.

CHALLENGE TASKS 8.1–8.2

8.1 A program to allow two players to compete in a game of chance is needed. The requirement is:

- The program must declare and instantiate an array containing 50 elements ready for storing integer values.

- The array must be populated by random values in the range 1 to 50.

- A second array is populated with values for the number of times each integer in the range 1 to 50 has been generated in the first array.

- Each player enters a number between 1 and 50.

- The player that has entered a value that has been generated the most number of times wins.

Note that your program must allow for the fact that the number of times a particular value has been generated can be zero.

8.2 Provide a design for a program that will do the following:

- Declare, instantiate and initialise an array containing six `String` values:

  - Each of these `String` values should have three characters.

  - The first should be a denary digit.

  - You should make sure that these denary digits are all different.

  - You should not have these denary digits in order.

  - Two examples are 6RT followed by 2AA.

- Take input of a search value that is one denary digit.

- Use a conditional loop that will compare the search value with the first character in successive array elements.

- If a match is found, output the last two characters for this element.

- The loop must terminate either if an element has been found with a first character that matches the search value or if the final element in the array has been checked.

Write the Java program to implement the design.

Bubble sort

There are many different sorting algorithms. The **bubble sort** is one of the simplest examples. A bubble sort algorithm steps through each element of an array and compares adjacent elements. If these elements are in the wrong order, the algorithm will swap them around. This is repeated until the list is sorted. The following is a pseudocode design for the bubble sort algorithm. The unsorted elements are stored in an array with name A. This is the only array used. The algorithm replaces unsorted element values with values sorted in order.

```
lastNumber ← lastIndex -1
FOR i ← 0 TO (lastIndex - 1) // the outer loop
    FOR count ← 0 TO lastNumber// the inner loop
        IF A[count] > A[count + 1]
// each element value is compared with the value
// in the next element of the array
        THEN
            Temp ← A[count]
            A[count] ← A[count + 1]
            A[count + 1] ← Temp
        ENDIF
    NEXT count // end inner loop
    lastNumber ← lastNumber – 1
NEXT i // end outer loop
```

You should note the use of a temporary value when you are exchanging two values. You cannot change two values in one action.

The following is an illustration of how the algorithm works for an array with 5 elements with index values 0 up to 4.

Consider the array A initially having the following values:

Index	0	1	2	3	4
Value	7	24	95	41	3

In the first iteration of the outer loop, the first two iterations of the inner loop leave the values unchanged. For the remaining inner loop iterations, the 95 successively moves to the last position in the array, leaving the values as:

Value	7	24	41	3	95

In the second iteration the 41 is moved to the last but one position to give:

Value	7	24	3	41	95

The following iterations give:

Value	7	3	24	41	95

then finally:

Value	3	7	24	41	95

The details for the progress of the algorithm are shown in Table 8.4.

Value for the outer loop counter i	Value for the variable LastNumber	Value for the inner loop counter Count	New value created for the variable Temp	New value created for an element in the array A	New value created for the next array element
0	3	0	no change	no change	no change
0	3	1	no change	no change	no change
0	3	2	95	41 for A[2]	95 for A[3]
0	3	3	95	3 for A[3]	95 for A[4]
1	2	0	no change	no change	no change
1	2	1	no change	no change	no change
1	2	2	41	3 for A[2]	41 for A[3]
2	1	0	no change	no change	no change
2	1	1	24	3 for A[1]	24 for A[2]
3	0	0	7	3 for A[0]	7 for A[1]

Table 8.4: The changes of the stored values as the bubble sort algorithm is executed

CHALLENGE TASK 8.3

Write a program for a bubble sort using the bubble sort pseudocode design we have just discussed. You can use the values for the elements of the array that are shown in the illustration of how the algorithm works.

8.5 Using combinations of arrays

So far, this chapter has focused on the use of a single array in a program. This single array can be described as a **one-dimensional (or 1D) array**. You could visualise this array either as a row of elements or as a column of elements. How the array is processed would be unaffected by this choice. However, there will be many applications that require a program that contains more than one array. In particular there can be multiple arrays with related content. Skills Focus 8.1 illustrates and explains some uses of multiple arrays.

SKILLS FOCUS 8.1

RELATED ARRAYS AND 2D ARRAYS

Let's consider how values for exam marks might be presented in a table. Here we are considering the sort of table that might be presented in a word-processing presentation or spreadsheet application. Table 8.5 is an example of the format that would be used.

	Maths	Computing
Candidate1	67	74
Candidate2	43	44
Candidate3	29	32
Candidate4	56	63

Table 8.5: Exam marks presented in a table

TIP

If you need to exchange values stored in two variables or in two array elements, you must begin by using a temporary variable to store one of the values.

KEY WORD

one-dimensional (or 1D) array: an array that can be visualised as either a row of values or as a column of values. It uses a single index to identify the elements.

CONTINUED

In a program, this data could be stored in three 1D arrays. Figure 8.1 shows how the three arrays would be related.

Index value applicable for each array	Values stored in candidate array	Values stored in mathsMark array	Values stored in computingMark array
0	Candidate1	67	74
1	Candidate2	43	44
2	Candidate3	29	32
3	Candidate4	56	63

Figure 8.1: Individual arrays with related index values

The following pseudocode design fragment shows how a program could take input to populate the three tables:

```
FOR i ← 0 TO 3
    OUTPUT "input candidate identifier"
    INPUT candidate[i]
    OUTPUT "input maths mark"
    INPUT mathsMark[i]
    OUTPUT "input computing mark"
    INPUT computingMark[i]
NEXT i
```

An alternative approach would be to use a **two-dimensional (2D) array**. This array could be named `examMarks`. Each element in a 2D array has an index for the row and an index for the column. Figure 8.2 shows how the index values identify an element in the array and the values stored. The values are the ones shown in Figure 8.1 for the exam marks.

KEY WORD

two-dimensional (or 2D) array: an array that can be visualised as a grid or a matrix.

Index values for 2D array examMarks	Col 0	Col 1
Row 0	0,0	0,1
Row 1	1,0	1,1
Row 2	2,0	2,1
Row 3	3,0	3,1

Values stored in 2D array examMarks	Col 0	Col 1
	67	74
	43	44
	29	32
	56	63

Figure 8.2: 2D arrays with related index values and values

Figure 8.2 shows that the first index identifies the row and the second index identifies the column. You can see that the 2D array has been populated with values so that the following rules apply:

- Row 0 contains marks for candidate1, row 1 marks for candidate2 and so on.
- Column 0 contains marks for the maths exam.
- Column 1 contains marks for the computing exam.

CONTINUED

Examples of the application of these rules are:

- The element examMarks[2,1] has the value for the mark of candidate 3 in the computing exam.

- The element examMarks[1,0] has the value for the mark of candidate 2 in the maths exam.

However, a 2D array does not contain any identifiers for what the values in the rows represent or for what the values in the columns represent. The programmer must decide how the rows and columns are to be used. Sensibly you would include a comment in a program to explain your choice.

The declaration for a 2D array uses the same format as shown for a 1D array in Section 8.2 except that the declaration must define the range of values for the row index followed by the range of values for the column index. In pseudocode the upper and lower bounds for row and column are defined. The generic version of the declaration statement for a 2D array using pseudocode can be represented as:

```
DECLARE <identifier> : ARRAY[<lr>:<ur>,<lc>:<uc>]

   OF <datatype>
```

For Java the declaration only defines the number of index values for row and for column with the assumption that the lower index value is zero in each case. The following would be the declaration statement for the examMarks array:

```
int[][] examMarks = new int[4][2];
```

When values are being entered into a 2D array or being read from a 2D array, a useful programming construct is a **nested iteration**.

The following is a pseudocode design for a nested iteration for entering marks into the array examMarks.

```
FOR rowCounter ← 0 TO 3
    FOR colCounter ← 0 TO 1
        candidate ← rowCounter + 1
        exam ← colCounter + 1
        OUTPUT "Enter mark for candidate " candidate
            "for exam " exam
        INPUT examMarks[rowCounter, colCounter]
    NEXT colCounter
NEXT rowCounter
```

Note that the code is written so that the user is not asked for a mark for candidate 0 or for exam 0.

Questions

1. Can you see a reason why the 2D array does not include a column to identify the candidate?

2. Can you see how nested for loops could be used when values were being entered into a 2D array?

> KEY WORD
>
> **nested iteration:** a programming construct used with 2D arrays where a For loop contains in the loop body a further For loop.

PRACTICE TASK 8.4

Implement a program that creates three 1D arrays. One array stores the names of candidates. The other two arrays store the marks that each candidate scores in two exams. You can store the values without asking for input. The program should conclude by outputting the scores for the candidates in the exams. The program need only allow there to be four candidates and two exams.

CHALLENGE TASK 8.4

Implement a program that uses a 2D array and a nested iteration for inputting values. The program is to store exam marks. The design for the nested iteration shown in Skills Focus 8.1 can be used. The program should provide as output the two marks for a candidate identified by an input.

SUMMARY

An array is an example of a data structure because it can store multiple values. In Java, an array is an object.
Elements of an array are identified by an index. The first index is zero.
An array has a length, which defines the number of elements.
There are two stages involved when an array is declared: the first stage creates the name, the second stage links the name to a memory location.
Many standard methods of solution, such as searching and sorting, use iteration with values stored in arrays.
In many applications, individual arrays are created where the content in one array is related to the content in another array. An alternative to using related 1D arrays is to use a 2D array.

END-OF-CHAPTER QUESTIONS

1 **a** Explain the term 'standard method of solution'.

    b Describe three examples of a standard method of solution.

    c Identify two programming constructs normally used for a standard method of solution.

2 The following represents data stored in an array with name `testArray`:

2	34	87	33	5
6	81	67	43	55
7	10	19	94	78

    a Identify the values stored in `testArray[2,1]` and in `testArray[1,2]`.

    b Write a pseudocode design that finds the maximum value in the array and outputs the value and its position in the array.

3 The following is an example of some data relating to household pets.

Type	Colour	Breed
Cat	White	Persian
Dog	Black and White	Husky
Dog	Grey	Poodle
Parrot	Blue and Red	Macaw

You need to design a program that will store this data in three related 1D arrays.
The system should be capable of being searched as follows:

- The user inputs an animal type and the program outputs all the data about animals of that type.
- The user inputs an animal type and breed and the program outputs any colours available.

4 Write the Java code for the design created for Question 3.

5 Provide a design and program code for the same problem as in Questions 3 and 4 but using a 2D array.

> Chapter 9

Designing algorithms

Introduction

Earlier chapters have shown examples of different ways of presenting a design. This chapter introduces an additional approach and considers the philosophy of top-down design. We will also discuss the context in which the design process takes place.

9.1 The context for the design process

As noted in Chapter 2, a program requiring only a few lines of code can be written using a programming language and implemented without any preliminary activities. However, for any sizeable program, the creation requires this sequence of activities:

1 problem definition
2 solution design
3 solution creation
4 solution testing.

We will first consider the development of a large-scale system by a team of computing professionals. The following will be features of the different stages:

- **Problem definition:** Problem definition will involve considering the feasibility of suggested solutions. There will be extensive analysis of what the system needs to do. The result will be a document that provides a detailed specification of the requirements for the system. This document will be passed on to the design team.

- **Solution design:** The design process will involve several stages. For example, the design of the interface and the design of the test plan will be separate activities. Some of these stages can occur at the same time.

- **Solution creation:** The creation of the final product will involve different parts of the system being allocated to different programmers.

- **Solution testing:** It is unlikely that the whole of the coding will be completed before testing begins. Instead, individual parts of the system will be tested as soon as the coding is complete for that part.

Unfortunately, when you are given a problem to solve, you cannot call upon a team of professionals. Instead, you will need to apply your own **computational thinking**. To do this, first look at the request you have received and make some decisions about what is required in a solution. The following is a sensible framework for applying computational thinking to achieve a detailed requirement specification. It suggests four steps, identifying what you should aim to do in each.

1 Identify the inputs and outputs that are involved in the scenario. The best approach is to start by deciding what outputs will be needed before considering what you will need as input data.

2 For each input, identify whether the problem requires repeated input. If it does, iteration will be needed. You may wish to identify the most appropriate loop to use.

3 For each output, identify the calculations required to produce the output value. You can ask yourself whether or not the calculations involve any decision making.

> **KEY WORD**
>
> **computational thinking:** a term that covers aspects such as abstraction, problem analysis, step-wise refinement, decomposition and algorithm design.

If they do, is there a need for IF statements in the program code? Similarly, will repeated calculations require some form of iteration?

4   Consider the sequence in which the various actions need to be completed: in particular, check that actions are correctly placed either inside, before or after any loop structures. As part of this, you can identify the variables you will be using.

DEMO TASK 9.1

You have been asked for a program that will receive input of 100 positive numbers.
The program will output the highest number and the sum of the numbers input.
You are required to produce a detailed analysis of the requirements.

Solution

Figure 9.1 shows the outcome of your computational thinking.

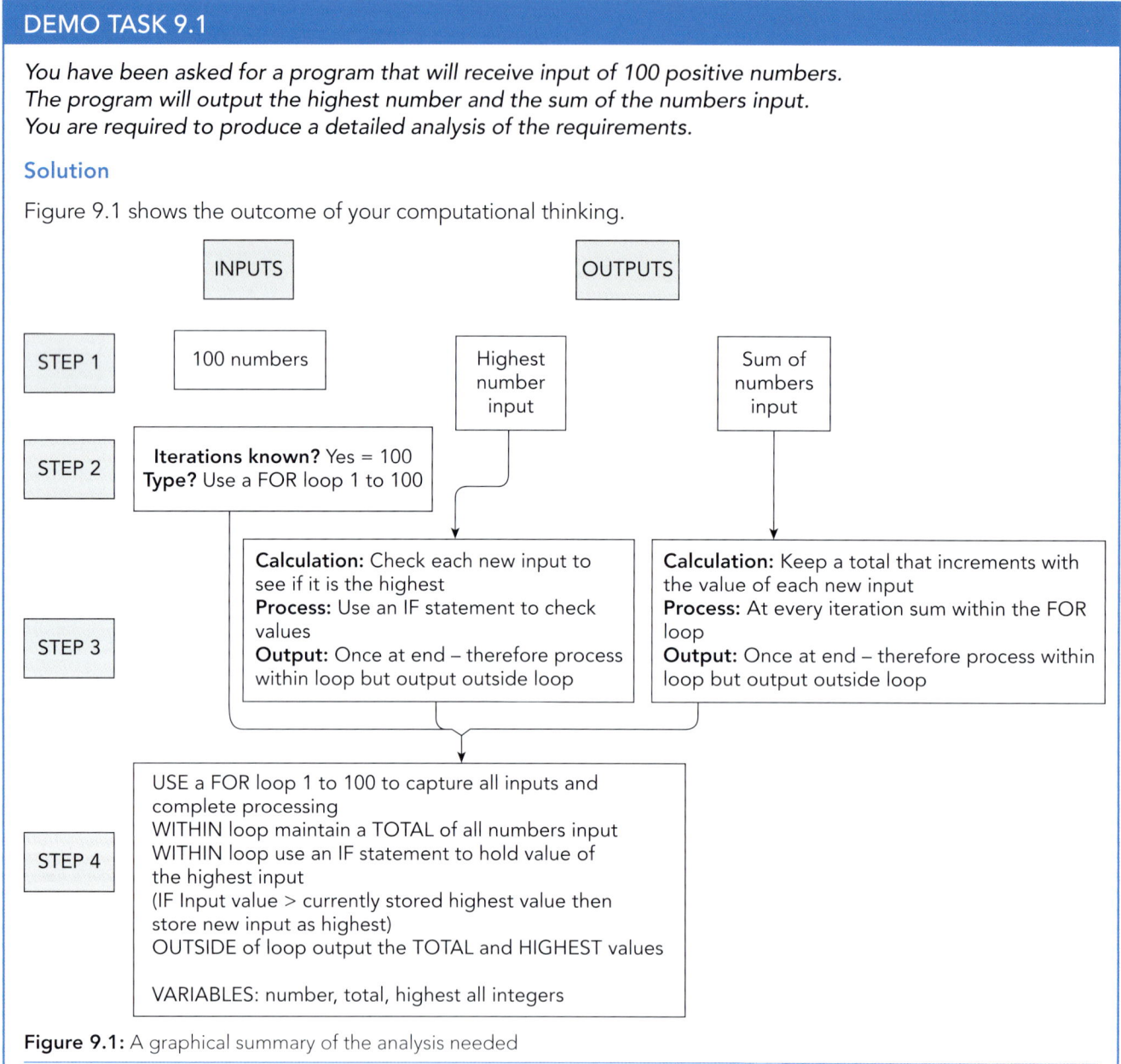

Figure 9.1: A graphical summary of the analysis needed

TIP

The boundary between when analysis is complete and design has begun is very often not clearly defined. You should not worry about this.

9.2 Top-down design

Top-down design is the approach that is used when designing an algorithm that is to be implemented using the constructs provided by an **imperative programming language**. The term implies that the starting point is an **abstraction** of the design. This will contain generalised statements about the design and very little, if any, detail.

There are two ways that top-down design can be put to use. One of these is the use of **step-wise refinement**. This is a process of increasingly adding detail. Typically, the starting point could be a Structured English description of an algorithm, as discussed in Chapter 2. Increasingly adding detail to this would eventually lead to a detailed design using pseudocode.

The alternative way of tackling top-down design is to use **decomposition**. Decomposition involves increasingly breaking down the design into smaller individual components. A **structure diagram** is a sensible choice of design technique to use here. This is explained and illustrated in Section 9.3.

9.3 Structure diagrams

Most structure diagrams can be mapped onto the structure shown in Figure 9.2. The presentation is consistent with the concept that a program algorithm typically involves the sequence:

1   the program takes input of some data
2   the program carries out the processing of data
3   the program outputs the result of the processing.

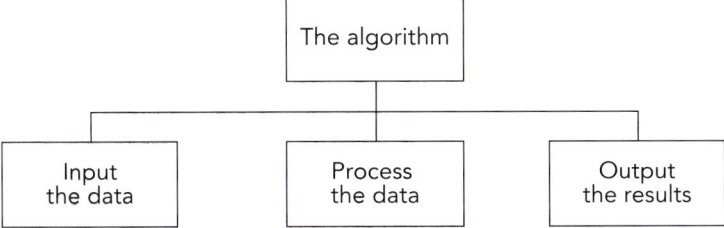

Figure 9.2: The conceptual starting point for a structure diagram

The Figure 9.2 diagram provides the following information:

- The name of the algorithm being designed is *The algorithm.*
- *The algorithm* consists of the three parts: *Input the data, Process the data* and *Output the results.*
- Each of these components represents one or more actions.
- There is a defined sequence from left to right.

There is no restriction on what each part of *The algorithm* represents. These could represent blocks of code in a single algorithm. Alternatively, they could represent modules that were intended to eventually be implemented as subroutines. Because a computer program can be described as a system, the approach is consistent with the concept of a system consisting of subsystems.

KEY WORDS

top-down design: a process where an initial abstraction is expanded to contain more detail, possibly by means of decomposition.

imperative or procedural language: a high-level programming language used to tell a computer what it has to do and how it should do it.

abstraction: an overview that contains the minimum necessary detail.

step-wise refinement: a process of increasingly adding detail to a design.

decomposition: a process of increasingly dividing a design into smaller components.

structure diagram: a hierarchical display of how a design can be broken down into smaller individual components.

DEMO TASK 9.2

Teachers in a school have to check student attendance in each lesson. A list is provided by the administration team, which is filled in by the teacher and then handed back to the administration team. If a student has not attended a lesson, the administration team contacts the parents by telephone text message or by email.

The school would like a system that could be accessed by teachers and by the administration team so that all of the actions needed could be handled by the system. Paper lists would no longer be needed.

Solution

You are going to create a structure diagram design using decomposition. The first stage of this is shown in Figure 9.3. The asterisk annotation indicates that a component requires an iteration.

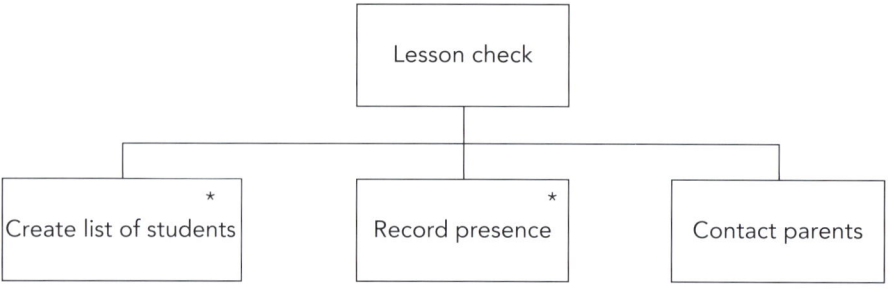

Figure 9.3: The first stage of the structure diagram design

You decide that the *Create list of students* component could be expanded, as could the *Contact parents* component. Figure 9.4 shows the expanded diagram.

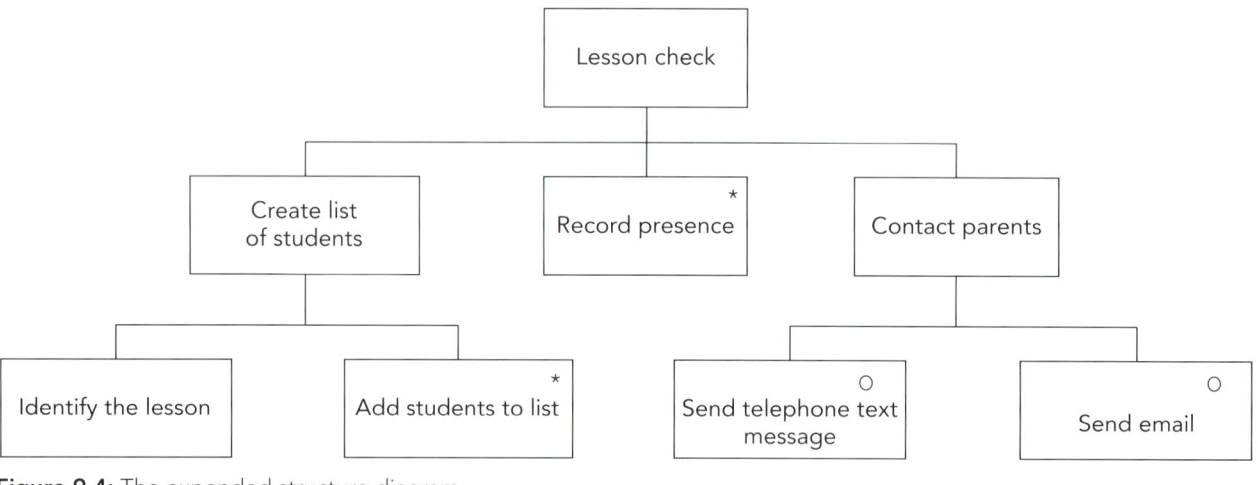

Figure 9.4: The expanded structure diagram

This shows the other option for annotation. The small circle indicates that there is a choice for how the parents are contacted. For the *Contact parents* component, only one of the lower components will be actioned.

Although it is possible to break the design down further, you decide that this will be sufficient.

Once a structure diagram has been created, there is an option to use step-wise refinement to complete the design for each component in the diagram.

DEMO TASK 9.3

You have decided to use the analysis in Demo Task 9.1 to document a design.

Solution

You begin by creating the structure diagram shown in Figure 9.5. You should note that here the design has been broken down in places to what would become individual program statements. This is an option you always have.

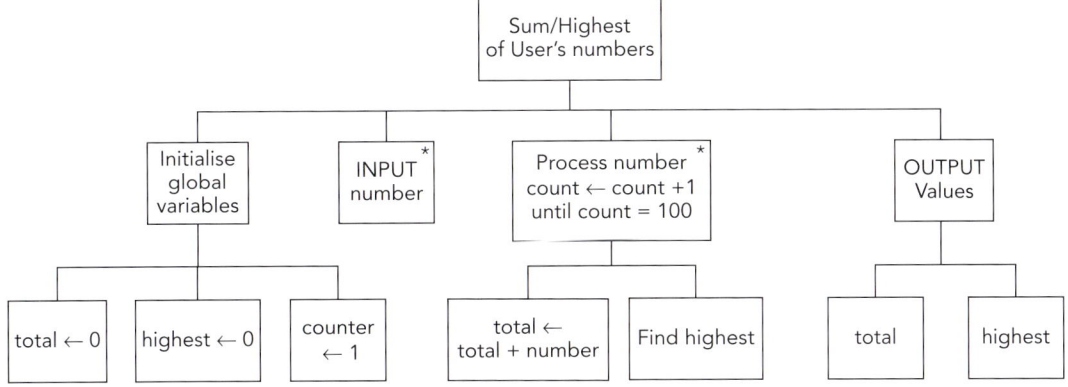

Figure 9.5: A structure diagram for the program analysed in Demo Task 9.1

The structure diagram can be used to create a flowchart or a pseudocode design. Flowchart 9.1 shows the flowchart version.

The pseudocode version of the design could be:

```
total ← 0
highest ← 0
FOR counter ← 1 TO 100
    INPUT number
    total ← total + number
    IF number > highest
      THEN
         highest ← number
    ENDIF
NEXT counter
OUTPUT total, highest
```

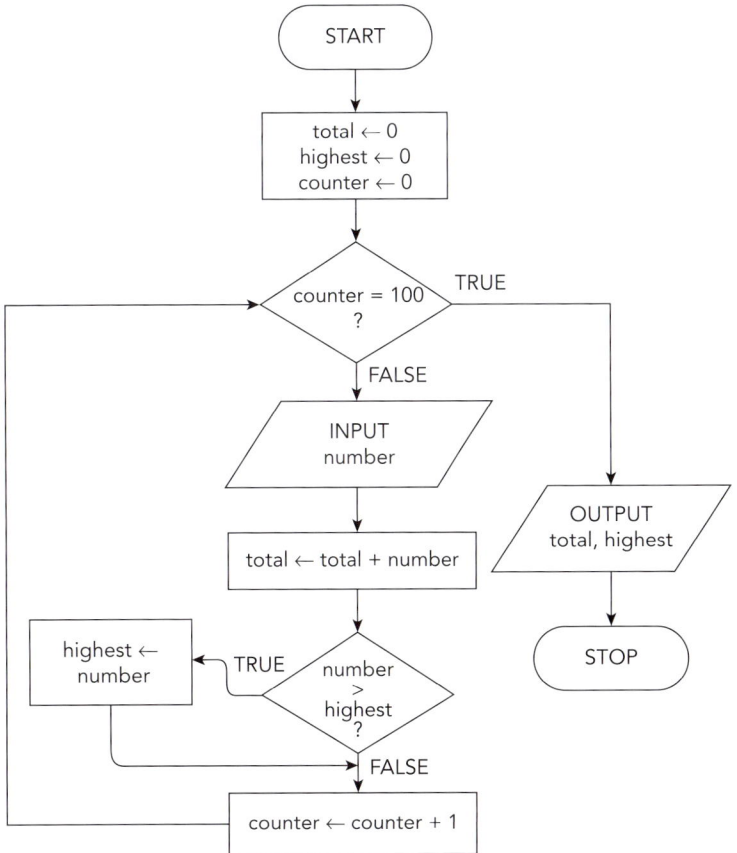

Flowchart 9.1: Flowchart version of the design

QUICK QUESTION

Are the flowchart and pseudocode designs in Demo Task 9.3 compatible?

PRACTICE TASKS 9.1–9.3

9.1 A librarian is producing a catalogue for a new set of books. The catalogue is to be stored in a computer system. A program is required that will receive input of the title and author for each book. The number of pages will also be entered. The librarian will enter a negative number for the number of pages to indicate that no more data will be entered.

Provide a structure diagram design for the program. Make sure the design includes initialisation of variables. When the diagram is complete, use it to create either a flowchart or pseudocode design.

Please note that the system being designed will need to permanently store an up-to-date version of the catalogue. This could be done by storing the data in one or more files. The need for permanent storage is inherent in many of the examples included in this chapter. The design you create here does not need to contain any details of the storage in files.

9.2 As an exercise in problem solving, consider the following scenario. The manager of a catering company has contacted a systems analyst and designer with a request for the design of a new system for the manager to use. The company provides a full service to customers for events such as weddings, anniversary celebrations or banquets. The new information system must store data that the manager can access when it is needed.

1   Create a list of likely details that the catering company would need to be obtained from the customer.

2   Create a list of likely details the company would need to store about the arrangements that the customer needs.

In answering the second question, you might wish to include a category relating to services from other suppliers that the catering company would need.

9.3 Figure 9.6 is a high-level structure diagram design for the system needed by the catering company as detailed in Practice Task 9.2. You can assume that the manager will be the only person using the system, including entering data and retrieving data.

Provide suitable content for the empty components in Figure 9.6 relating to data stored and retrieved.

CONTINUED

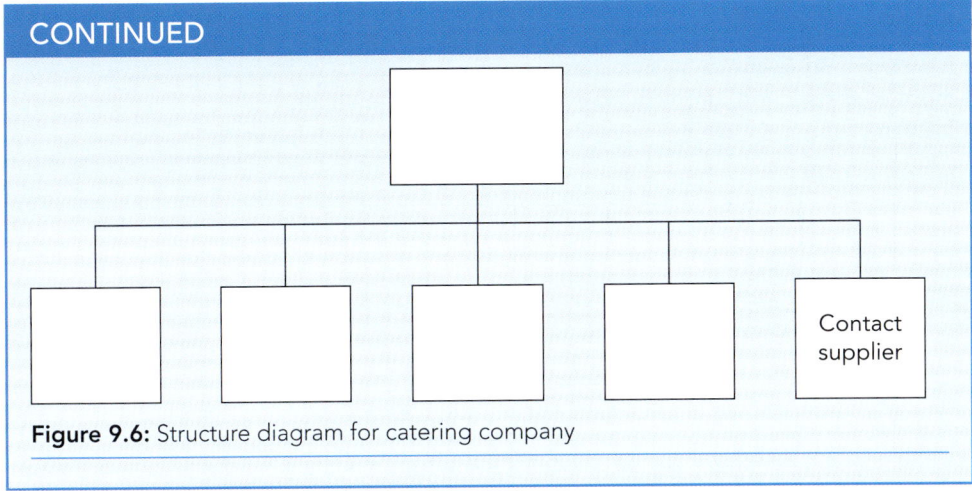

Figure 9.6: Structure diagram for catering company

CHALLENGE TASKS 9.1–9.2

9.1 A system is designed to collect monthly rainfall data in millimetres from weather stations around the UK. It then has to output a monthly rainfall figure for each location and the average for the whole country. Unfortunately, not all weather stations will fill in their data each day. Draw a structure diagram design that breaks this system down into subsystems.

Note that there is more than one correct design that would match this requirement.

9.2 Consider the following scenario.

An aircraft charter company that owns a fleet of aircraft and employs several pilots runs charter flights for its customers. Each individual flight may or may not have a co-pilot in addition to the pilot; however, it will always have three cabin crew. Each flight may land at an intermediate airport before reaching its destination.

The company needs a system for storing information about the flights that have been arranged.

a   Construct a list of the data that would be needed to be stored in the system.

b   Draw a structure diagram to represent this system. The diagram should include some lower-level detail – for example, to allow for the possibility of a co-pilot or for the use of an intermediate airport.

SUMMARY

Computational thinking incorporates abstraction, problem analysis, step-wise refinement, decomposition and algorithm design.
Top-down design starts with an abstraction.
Step-wise refinement and decomposition are alternative approaches to top-down design.
A structure diagram is a useful tool for presenting decomposition.
A structure diagram has a hierarchical structure and it defines a sequence, which is represented as left to right.
A structure diagram can include annotations indicating iteration or optional components
A structure diagram can be used to create a flowchart or a pseudocode design.

END-OF-CHAPTER QUESTIONS

1 A weather app works by using location data entered by the user, either a new town or a chosen town from a previously saved location. The weather app will then output the day's forecast either as a visual map or as a table of temperatures, wind speeds and weather icons. The structure diagram below is a partially completed design. Copy out the diagram and enter names for the components currently lacking information.

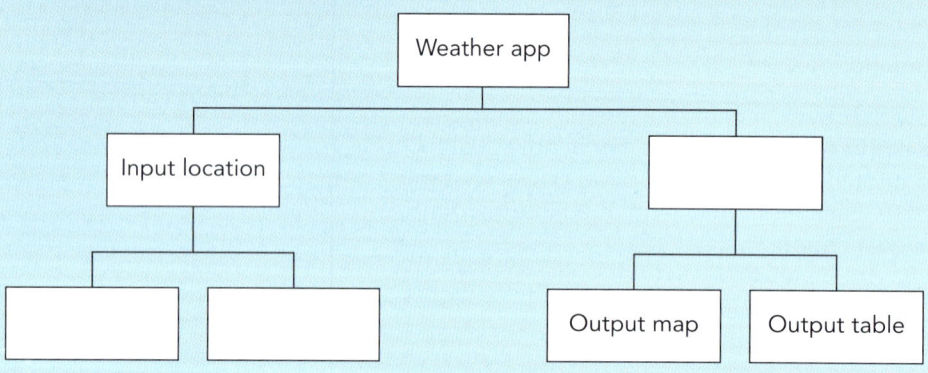

2 A health centre needs a system that can be used by the receptionist to store information about all activities that need arranging or have been arranged. The following are some typical activities:
- Booking appointments with doctors.
- Booking appointments with nurses for vaccination or for taking blood samples.
- Booking an ambulance.
- Booking a hospital appointment.
- Receiving telephone queries.
- Contacting patients.

Construct a structure diagram that would be a suitable design for the system.

Checking inputs

Introduction

Data needs to be accurate when it is stored in a computer system or is being processed by a computer program. This is known as data integrity. Data only has integrity if it is accurate and up-to-date. The most likely cause of incorrect data is that it is initially entered incorrectly. The focus in this chapter is on methods that can be used to minimise the chances of inaccurate data being input into a system.

10.1 Validation and verification

Validation is checking that data matches specified criteria. There might be a number of validation checks applied to a single data item. For example, a telephone number might be checked to see that there are only numeric digits being entered and that the number of digits is correct.

The important point about validation is that it does not guarantee **data integrity**. If you were entering your name into a system, you would be unlikely to spell it incorrectly. However, if someone else were entering your name, then that person might misspell it. The data entered is then incorrect but it would not fail a validation check because it would still be recognised as a name.

As we will see in Section 10.3, validation can be handled by the inclusion of appropriate code in a program. If invalid data is detected there can be a request for a valid entry.

Verification is a check that data entered is what was intended to be entered. One scenario for the use of verification is when data is being transmitted from a sending system to a receiving system.

10.2 Categories of validation

Before looking at how a program can include suitable code for carrying out validation, it is useful to consider how validation can be categorised. Table 10.1 contains a list of the commonly used categories. For each of these, there is a definition and a discussion about likely uses.

Category	Definition	Possible uses
Length check	Checks how many characters are input.	In principle, this could be used for numeric data but the use will almost certainly relate to the input of a string of characters. One use is to check that a telephone number has the correct number of digits. Often an organisation will have rules about the maximum number of characters allowed for names of departments, for example, or perhaps the minimum number of characters allowed for a password.

(continued)

TIP

It is easy to get confused about what validation is and what verification is. It is worth your while to spend some time memorising the meanings of these words.

KEY WORDS

validation: the process of programming a system to automatically check that data satisfies a set of specified input criteria; for example, passwords must be longer than six characters.

data integrity: the term used to describe data that is accurate and up-to-date.

verification: a check that the data entered is what was intended to be entered.

Category	Definition	Possible uses
Range check	Checks against defined lowest and highest values.	Although character data could be tested – for example, to check if a character was an upper case letter – this validation would normally be used on numeric data. Usually, the aim would be to detect impossible values. Examples include: • an age input as 125 • a height for a person entered as 3 metres • an exam mark entered as 90 when the highest mark possible for the paper is 75 • a negative value input when only a positive value would make sense.
Presence check	Checks whether a value has been input.	This could be used to check that a user had not forgotten to enter a value before hitting the enter key. More likely, this would be a check when a menu or a tick box was presented to a user and the user had failed to indicate a choice.
Type check	Checks that the input value has the correct data type.	This could be used to check whether a user had input a word rather than a number. It could also be used when a name was being entered so that only alphabetic characters would be acceptable.
Format check	Checks that the sequence of input characters matches a defined pattern.	The usual example for this check is when entering a date. However, there are also examples where an input string should have spaces between separate parts or hyphens. For example, an ISBN is usually printed with four hyphens.

Table 10.1: Categories of validation

As a general point, it can be noted that if validation is not used, the only effect may be the minor irritation of receiving an error message from the compiler. For example, if you have declared a variable as int then you try to input a String variable using the same variable name, you get a compiler error message like this:

novalid.java:12: error: incompatible types: String cannot be converted to int

```
yourName = myObj.nextLine();
           ^
```

1 error

The effects of a lack of validation can be much more severe if the compiler cannot see any error in the input coding and the program code is allowed to be executed with incorrect data.

10.3 Programming validation

We can now look at how validation checks can be programmed for each of the categories listed in Table 10.1.

Length check validation

DEMO TASK 10.1

You have been asked to create an interface for a system. The system interface will ask the user to input a password of six or more characters. The system should only accept passwords with six or more characters. It should produce an error message if the submitted password contains five or fewer characters.

Solution

You decide that there is no sense in allowing the error message to cause an exit from the program. Instead, a loop structure is required that informs the user of the problem and asks for a resubmission of a password, but this time one with an acceptable number of characters.

You create the flowchart shown in Flowchart 10.1 as a suitable design to define the overall logic for the **length check**.

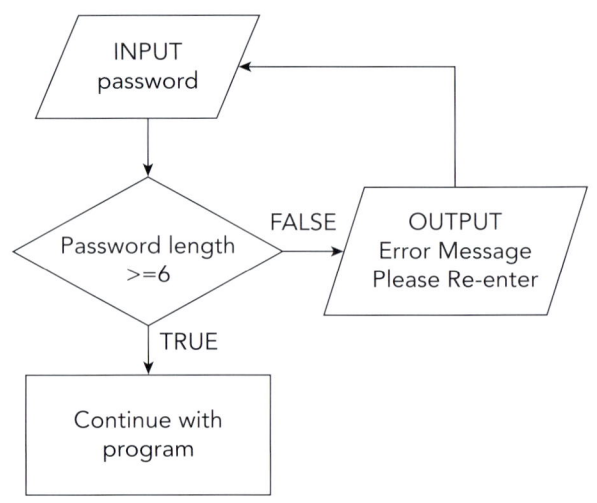

Flowchart 10.1: A flowchart for a length check

You can program this with a While loop to make sure that the user does eventually supply a sensible password. The Java code you choose for the validation check could be:

```java
sensiblePW = false
while (sensiblePW == false)
{
    System.out.println("Please enter your choice of a
        password");
    Scanner PWInput = new Scanner(System.in);
    yourPW = PWInput.nextLine();
    if (yourPW.length() >= 6) sensiblePW = true;
    else
    System.out.println("Try again with at least six characters");
}
```

> **KEY WORD**
>
> **length check:** a check that an input has the correct number of characters.

Range check validation

A program for a **range check** validation could be created with the same structure as the one suggested for the length check. The following pseudocode design uses a slightly different approach. The program is to receive an integer value that represents the day in a month. Validation is needed to check that the value entered is in the range 1 to 31.

```
INPUT day

WHILE day < 1 OR day > 31 DO

    OUTPUT "The value should be in the range 1 - 31
    please try again"

    INPUT day

ENDWHILE
```

Code snippet 10.1

Presence check validation

The following program demonstrates how a **presence check** can use the methods associated with Scanner. This particular program first checks to see if anything has been entered, then checks if the value supplied for a percentage is in the range 0–100.

KEY WORD

presence check: a check that there has been a value entered.

```java
import java.util.Scanner;
/*
checks for presence then checks range
*/
class PresenceCheck {
public static void main(String[] args) {
boolean sensibleValue = false;
int percent = 0;
String stringIn = "";
while (sensibleValue == false)
{
    System.out.println("Please enter the percentage");
    Scanner theInput = new Scanner(System.in);
    stringIn = theInput.nextLine();
    Scanner strObj = new Scanner(stringIn);
    if (strObj.hasNext() == false)
        {
        System.out.println("no value entered");
        continue;
        }
    percent = strObj.nextInt();
    if (!(percent >= 0 && percent <= 100))
        {
        System.out.println("not in range, try again");
        continue;
        }
    if (percent >= 0 && percent <= 100)
        sensibleValue = true;
}
System.out.println("Percentage is " + percent);
}}
```

Code snippet 10.2

The code highlighted in brown in Code snippet 10.2 is where the Scanner methods are used. These are explained in Table 10.2.

Java code	Explanation
`Scanner theInput = new Scanner(System.in);`	This is the normal use of `Scanner` to get input.
`stringIn = theInput.nextLine();`	This converts the input into a `String` value stored in the variable `stringIn`.
`Scanner strObj = new Scanner(stringIn);`	This shows the normal use of `Scanner` but applied to the string held in `stringIn` rather than directly to the input.
`if (strObj.hasNext() == false)`	This is the check on the presence; `hasNext()` is a Boolean method that returns true if there is any content.
`percent = strObj.nextInt();`	This is where the object created by `Scanner` is interpreted as an integer and stored in the variable `percent`.

Table 10.2: Scanner methods explained

Note the use of the `continue` statement in this code. The `continue` statement causes the current iteration to stop and the next iteration to start (dependent, of course, on the loop condition still being met).

PRACTICE TASK 10.3

Implement the Java code shown in Code snippet 10.2 for a combined presence check and range check.

Type check validation

The simplest use of a **type check** validation is when there is a single input of a value. For example, if an integer value is needed, Flowchart 10.2 can be used as a design.

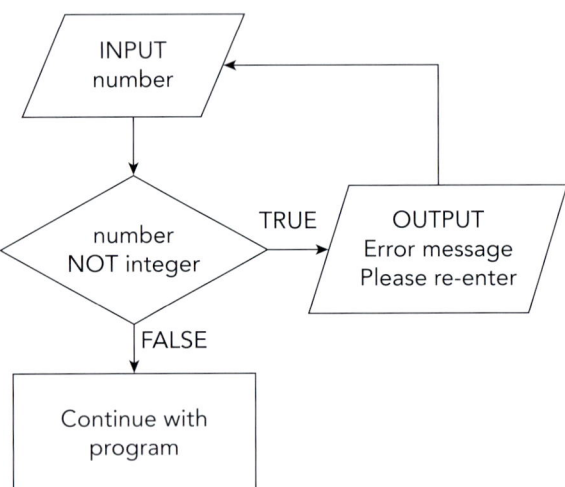

Flowchart 10.2: Flowchart for a type check

The following Java code for the type check uses the Scanner method `hasNextInt`. This returns a value true if an integer is entered. Note also in this code the use of the break statement, which causes an immediate exit from the loop.

```
sensibleInt = false
while (!sensibleInt)
{
    System.out.println("Please enter an integer value");
    Scanner intObj = new Scanner(System.in);
    if intObj.hasNextInt()
        {
        intVal = intObj.nextInt();
        break;
        }
    System.out.println("Not an integer");
}
```

Code snippet 10.3

PRACTICE TASK 10.4

Implement a Java program incorporating the code in Code snippet 10.3 to check for an integer input.

A more difficult error to guard against is if a string being entered contains one or more characters of the wrong type. Some of the sophisticated and complicated facilities offered by Java are briefly considered under the heading of Format checking which comes after this further example of type checking.

In the meantime, we can consider a check to ensure that when a name is entered, there are no numeric digits included by mistake. The following Java program could be used for this.

The statements highlighted in brown successively check each character in the variable.

```java
import java.util.Scanner;
/*
check for illegal number in a string
 */
class numberInString {
public static void main(String[] args) {
String yourName = "";
boolean needName = true;
int len = 0;
boolean problem = false;
while (needName)
    {
    problem = false;
    System.out.println("Please enter your name");
    Scanner strObj = new Scanner(System.in);
    yourName = strObj.nextLine();
    len = yourName.length();
    for (int i = 0; i <= (len - 1); i++)
        {
        if (Character.isDigit(yourName.charAt(i)))
        problem = true;
        }
    if (problem) System.out.println("no numbers allowed");
    else
    {
        needName = false;
        System.out.println("Name accepted");
        }
    }
}}
```

Code snippet 10.4

PRACTICE TASK 10.5

Iimplement a Java program incorporating the code in Code snippet 10.4 to check for an integer included in a string.

QUICK QUESTION

The code supplied for the program to be used for Practice Task 10.5 is not efficient. More processing is possible than there needs to be. Can you see what the problem is and how it might be removed?

Format check validation

Java offers an extensive range of techniques concerning pattern matching. An example is presented here which illustrates the use of a **format check**. The following program needs an input of four numeric digits. The special features of the program are highlighted in brown. These are discussed in Table 10.3.

```
import java.util.Scanner;
import java.util.regex.Pattern;
import java.util.regex.Matcher;
/*
 FormatCheck
 */
class formatCheck {
public static void main(String[] args) {
    String inputStr = "";
    System.out.println("Please enter four digits");
    Scanner strObj = new Scanner(System.in);
    inputStr = strObj.nextLine();
    Pattern myPat = Pattern.compile("[0-9]{4}");
    Matcher myMat = myPat.matcher(inputStr);
    if (myMat.matches())
        System.out.println( inputStr + " correct format");
    else
        System.out.println("Not 4 digits");
}}
```

Java code	Explanation

`import java.util.Scanner;`
`import java.util.regex.Pattern;`
`import java.util.regex.Matcher;`
	These are the methods that must be made available for use in the program.

`Pattern myPat = Pattern.` `    compile("[0-9]{4}");`	This coding tells the computer what format the required input needs to be in. The first part is declaring a variable `myPat` of type `Pattern`. `Pattern.compile` is a function. The argument of this function is an example of a **regular expression**. A regular expression definition is enclosed in double quotes. In this example: • `[0-9]` says that any numeric digit is allowed. • `{4}` says that there must be four numeric digits.

(continued)

Java code	Explanation
`Matcher myMat = myPat.` `    matcher(inputStr);`	This statement is comparing the characters in the `String` variable `inputStr` with those defined in the `Pattern`, `myPat`. The result of the comparison is stored in the variable `myMat` of type `Matcher`.
`if (myMat.matches())`	This is the test of the comparison. If the input string did match the defined pattern, this will return true. If the input string did not match the defined pattern, this will return false.

Table 10.3: Explanations of the use of `Pattern` and `Matcher`

10.4 Verification

Verification concerns good practice when entering data. Programs are used to support verification, but the focus here will be on the aims of the possible approaches. The following are examples:

1   The simplest approach is for the user to look at the data that is on the screen following keyboard input. If there is an error, the user can then change their input before they hit the Enter key to complete the input process.

2   In order to encourage the user to check the data, the system could echo what has just been entered.

3   If the user is entering a password, it would not be good practice to echo this on a screen. Indeed, it is almost universal practice for there to be no display of the password when it is being entered. Instead, it is normal for a repeat entry to be required for the system to check that the two entries are identical.

4   When security is a concern, the issue is not whether the input data is what was intended to be input. The issue is whether the person inputting the data is fraudulently using someone else's identity. In this case, supporting security checks are carried out. The usual approach is to ask an additional question that only the genuine person should be able to answer correctly.

5   A special case is the use of a CAPTCHA image to ensure that a person, not a computer, is accessing the system.

10.5 Check digit

A **check digit** is a validation method which can only be used with a particular form of data. It can be used whenever the data consists of a character string of defined length containing numeric digits. A check digit can be calculated when a value is about to be stored. It is then appended as a single character to the original data string. It is calculated again when the data is accessed. If the two calculations do not match, this is evidence that what was originally input for storage was not what was intended.

The check digit method is normally used when the data to be input has a relatively small number of numeric digits. The usual examples quoted are the ISBN and the bar code. We will consider the ISBN here. Many years ago, an ISBN consisted of ten digits. In the modern version, there are 13 digits.

Flowchart 10.3 shows a flowchart for an ISBN-13 check digit validation. The process to find the check digit is as follows:

1   The 13th digit is removed as this is the check digit.

2   All the other numbers are assigned a 1 or 3, alternating from 1. These numbers are used as multipliers for their corresponding digits.

3   All the products are added together and the remainder after dividing this by 10 is found.

4   The remainder is then subtracted from 10. This should match the check digit if the ISBN number is correct.

For example, the ISBN of this book is 978-1-108-91007-1.

1   The final 1 is removed.

2

ISBN digit	9	7	8	1	1	0	8	9	1	0	0	7
Multiplier	1	3	1	3	1	3	1	3	1	3	1	3
Product	9	21	8	3	1	0	8	27	1	0	0	21

3   Total = 99. The remainder after dividing this by 10 = 9.

4   10 − 9 = 1.

The check digit is 1. Compare this with the 13th digit in the ISBN of this book.

KEY WORD

check digit: a value that is calculated from the numeric digits in a code used for identification and is then added to the code as an additional digit.

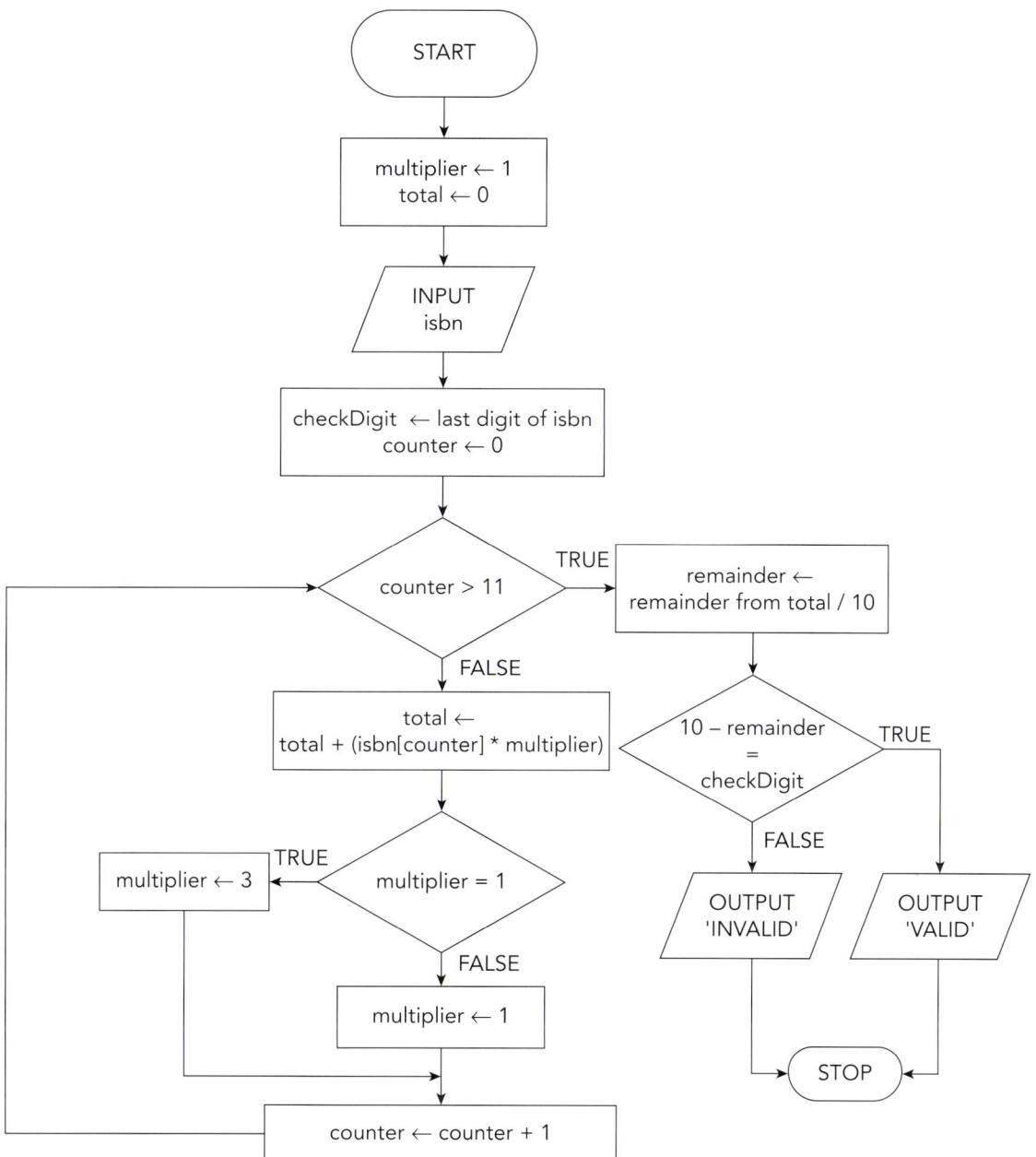

Flowchart 10.3: A flowchart design for an ISBN check

SUMMARY

Data integrity requires data to be accurate and up-to-date.
A major cause of a lack of data integrity is the incorrect input of the data.
Data validation checks that the input data conforms with defined criteria.
Data validation can refer to a length check, a range check, a presence check, a type check or a format check.
Data verification is aimed at ensuring that data entered is in fact what was intended to be entered.
A check digit can be calculated from the data and added as an extra digit if the data consists only of numeric digits.

END-OF-CHAPTER QUESTIONS

1 Explain the difference between a range validation check and a length validation check.

2 Explain three different methods of verification.

3 Write the code for a function that will validate a username and password entered by a user of a computer system.

   The inputs must be validated against the following criteria:

   • The username must consist of only lower case letters and hyphens.

   • The password must start with an upper case P and then consist of 6 integers, e.g. P537917.

   The function will be passed the username and password as arguments and will return a boolean value to indicate if the inputs match the validation criteria.

4 Produce a flowchart or pseudocode for a system that makes use of the function in End-of-Chapter Task 3. If the validation is completed successfully, the system will output 'Welcome'. If the validation check is failed once, the system will output 'Error. Please try again'. If the validation check fails twice, the system will output 'Locked Out' and exit the program.

Chapter 11

Testing

IN THIS CHAPTER YOU WILL:

- learn about the three categories of error: syntax, runtime and logical errors

- learn about the causes of these errors

- understand the place of testing in the different stages of the development of a large-scale software-based system

- learn how to use normal, abnormal, extreme and boundary test data

- understand how a debugger can help locate errors

- perform a dry-run using a trace table to help in locating errors.

Introduction

There are several different types of errors that you will come across in programming. This chapter defines how the errors that can be present in a program can be categorised. We will look at methods of dealing with errors in your programs. In addition, there is a discussion of testing in the context of large-scale commercial systems.

11.1 Categories of error

Syntax errors

The category of error that you will first encounter as a programmer is the **syntax error**. If you have not already encountered a syntax error, this must be because you haven't attempted to run many programs. We all write code containing syntax errors.

For Java, the most common syntax errors are:

- A failure to put a semi-colon to conclude a statement.

- Not including properly paired braces.

- Not using the correct case for a component in the code.

The good thing about syntax errors is that the compiler tells you about them when you submit your program for compilation. Most of the time, the error messages that you get from the compiler are easy to understand. However, the number of error messages sometimes makes the problem appear to be worse than it really is. This is because one error can lead to more than one error message.

Runtime errors

If all of the syntax errors have been corrected, you do not get congratulations from the compiler. Instead, the absence of an error message is the sign that you can attempt to run the program. Unfortunately, the absence of any syntax errors does not necessarily mean that the program will run. It might start to run, but then you could meet the second kind of error. A program will halt if there is a **runtime error**. When the program fails, there will be a message output to identify the cause of this.

The most common causes of runtime errors are:

- Attempting one iteration too many in a loop.

- Using a value for the index of an array that is not in the defined range.

- Carrying out a calculation that results in a value being assigned to a variable which is too large for the memory storage space allocated for the variable. This is called an **overflow error**.

- Attempting a division when the variable being used to divide has a value of zero.

The characteristic of a runtime error is that it arises from a value supplied to the program when it is running. Many of the methods for checking inputs that were discussed in Chapter 10 are aimed at minimising the chances of runtime errors.

KEY WORDS

syntax error: an error that occurs when the code you have written does not conform to the rules defined for the language.

runtime error: an error that occurs when the values being used by the program lead to the program being unable to perform an action.

overflow error: an error caused by a calculation producing a value for a variable that is too large for the storage space allocated in memory for the variable.

TIP

If you get an error message from the compiler that you do not understand, you can get explanations if you copy and paste the message into a search engine.

TIP

Division by zero can be avoided by using a check that the dividing value is not zero immediately before the statement using division. A simple error message can then be output if needed.

Logical errors

Even when the compiler reports no syntax errors and the program has run to completion, there is no guarantee that the program is error free. A program may have the third kind of error: a **logical error**.

Typical causes of logical errors are:

- Incorrectly formulated arithmetic expressions in an assignment statement.

- The use of the wrong arithmetic operator.

- Incorrect use of Boolean logic.

You will only be able to realise that a program contains logic errors if you have an expectation of what the program should produce as output for given input values. This is where testing becomes vitally important.

KEY WORD

logical error: an error which is due to the program code having faulty logic; this allows the program to run to completion but with output that is not what should have been produced.

11.2 The testing framework

The focus here is on the use of testing in large-scale development projects. All such projects require extensive testing. However, the need is even greater if a safety-critical system is being developed, such as an air traffic control system.

Unfortunately, failures to detect errors can lead to catastrophic results. At the time of writing, a major aircraft manufacturer has had a fleet of airliners grounded following two fatal crashes. The software for an automatic system for correction of the angle of tilt of the aircraft has been identified as a cause of the problem. The destruction of the unmanned *Ariane 5* space rocket, due to the failure of untested code, is one of the most costly examples ever. The financial losses were measured in billions of dollars. An article published in *The New York Times* magazine in December 1996 has information about the software error that caused the disaster. You will be able to find the article if you search online for 'Ariane disaster New York Times'.

The context for large-scale system development is that there is a client requiring a new system and a development team expecting to deliver the product. In Chapter 9, we looked at the following sequence as a framework for the development of such a system:

1   problem definition
2   solution design
3   solution creation
4   solution testing

It was mentioned there that part of the solution design stage should be the creation of a test plan. A test plan must contain a variety of individual tests. You should choose a range of tests to test all possible routes through the program code. For each test, you will need defined input values. These can be classified as shown in Table 11.1. The example data given in Table 11.1 could be used to test a program that receives values for percentage marks in an exam as input.

Category of data	Description	Example data
Normal data	Input data that is acceptable and expected not to cause any error.	A percentage mark of 25.
Abnormal data	Input data that should be rejected because it has an impossible value.	A percentage mark of 134.
Extreme data	Input data that is on the limit of the range of normal data values.	A percentage mark of either 0 or 100.
Boundary data	Input data with the largest or smallest acceptable values and the corresponding smallest or largest values that should be rejected.	A percentage mark of 0 or 100 are acceptable values. A percentage mark of 101 or −1 should be rejected.

Table 11.1: The four categories of test input values

The four categories have different aims:

1 Testing with **normal data** is aimed at checking for logical errors. There should be definitions of the expected output for the given input values.

2 Testing with **abnormal data** is aimed at checking that the appropriate validation rules have been coded in the program. The program should provide an immediate message.

3 Testing with **extreme data** is aimed at checking program logic.

4 Testing with **boundary data** is aimed at checking that loop constructs and array indexing have been correctly coded.

Testing will begin as soon as a part of the system has been coded. During the solution creation stage, there will be initially only coding activity. Later on, testing will be happening as well as coding. When all of the coding has been completed, the solution testing stage can begin. The first aim of this is to test that all of the component parts of the system are still performing properly when assembled into the complete system.

At the beginning of the solution testing stage, the development team will still be carrying out the testing on their own systems. This is categorised as **alpha testing**. When the results of alpha testing indicate that the system is functioning properly the product will progress to **beta testing**. This is when the product that has been developed is installed on the systems where it is intended to be used. The development team might still be carrying out the beta testing. Alternatively, it will be normal users of the new system who test it just by using it.

PRACTICE TASK 11.1

The following program records the highest temperature recorded for each month. It takes a daily input of a maximum temperature value for that day. The value is input along with the month number. The program then checks to see if this value is higher than the value already stored for that month.

KEY WORDS

normal data: values used for testing that should produce sensible output.

abnormal data: values used for testing that the program should detect as being impossible values.

extreme data: values used for testing that are on the limit of the range of normal values.

boundary data: values used for testing that are just inside or just outside the range limits for normal data.

alpha testing: the stage of testing when the development team are testing using their own systems.

beta testing: the stage of testing when the developed product has been installed on the client's systems.

CONTINUED

```java
import java.util.Scanner;
class practiceTest {
public static void main(String[] args) {

    int[] monthTemperature = new int[12];
    int monthNumber = 0;
    int temperatureValue = 0;
    for (int i = 0; i <= 11; i++)
        {
        monthTemperature[i] = -25;
        }
    System.out.println("Which month number?");
    Scanner myObj = new Scanner(System.in);
    monthNumber = myObj.nextInt();
    System.out.println("What is the temperature?");
    Scanner myObj1 = new Scanner(System.in);
    temperatureValue = myObj1.nextInt();
    if (temperatureValue > monthTemperature[monthNumber])
        monthTemperature[monthNumber] = temperatureValue;
    System.out.println("Thank you for the reading");
}}
```

Can you suggest:

- a pair of input values representing normal data

- a pair of input values representing abnormal data

- a set of input values representing boundary data?

11.3 Debugging

Debugging is the process of locating and correcting runtime or logical errors in a program. For a runtime error, the error message will often indicate how the error can be corrected. In other cases, it will be necessary to know the values stored for variables in use prior to where in the program the error is taking place. For a logical error, the programmer may only have a suspicion of where the source of the error is in the program. There are four possible methods to use for finding out the values stored for variables:

1   A memory dump. This is where a representation of the values stored in memory is produced. This is a method traditionally used by experienced programmers, but it is rarely used nowadays.

2   Inserting statements that output the values of variables at selective points in the code. Again, this is a traditional approach and would now only be used occasionally.

3 Using debugging software custom-built for the programming language being used.

4 Dry running using a trace table. This can be used with a program design or with program code. It can be a useful approach for becoming familiar with what a design or program is doing when this was produced by someone else.

Sections 11.4 and 11.5 will consider methods 3 and 4 on this list.

11.4 Using a debugger

A **debugger** is specialised software aimed at locating errors in a program. It allows execution of a program in a controlled manner. The main features of a debugger are:

- The lines of the code are numbered.

- A program can be executed line-by-line.

- A **breakpoint** can be inserted to halt execution at a specified line of the code.

- The values of variables can be displayed at any stage.

If you are using an IDE for creating your programs, it should have an option for running a debugger. The alternative is to use the built-in debugger provided by Java.

In order to use the Java debugger in this simple form, it is necessary to comment out all of the input and output lines of code. (If you do not do this the debugger does strange things.) Also, whenever the last line in a program is an output statement, an additional line of code should be inserted after this. Code snippet 11.1 shows an edited version of the code in Practice Task 11.1. The class name has been changed. There are also some indentation changes but the logic of the program is unchanged.

(You should note that the line numbers have only been added here to help understanding. You would never put line numbers in a version of a program you intended to run.)

```
1.  import java.util.Scanner;
2.  class pracTest {
3.  public static void main(String[] args) {
4.
5.  int[] monthTemperature = new int[12];
6.  int monthNumber = 0;
7.  int temperatureValue = 0;
8.  for (int i = 0; i <= 11; i++)
9.  {
10.     monthTemperature[i] = -25;
11.     }
12.     //System.out.println("Which month number?");
13.     //Scanner myObj = new Scanner(System.in);
14.     //monthNumber = myObj.nextInt();
15.     //System.out.println("What is the temperature?");
16.     //Scanner myObj1 = new Scanner(System.in);
17.     //temperatureValue = myObj1.nextInt();
18.     if (temperatureValue >
        monthTemperature[monthNumber])
            monthTemperature[monthNumber] = temperatureValue;
```

KEY WORDS

debugger: specialised software used to locate errors in a program.

breakpoint: a facility used in a debugger to identify a line of the code where execution should stop.

```
19.    //System.out.println("Thank-you for the reading");
20.    temperatureValue = temperatureValue + 1;
21.    }
22.    }
```

Code snippet 11.1

Table 11.2 shows the debugger being used to examine Code snippet 11.1. In the first column (Screen display) the parts that are highlighted in brown are those that you enter at the keyboard. The text in black is what is displayed after you have pressed the Enter key.

Screen display (abbreviated in places)	Explanation
C:\Users\daved\Documents\Java progs>javac -g pracTest.java C:\Users\daved\Documents\Java progs>jdb pracTest7 Initializing jdb ...	Using the -g argument ensures that you will get the most information from the debugger. The jdb command starts the debugger.
> stop in pracTest.main Deferring breakpoint pracTest.main. It will be set after the class is loaded. > stop at pracTest: 18 Deferring breakpoint practiceTest:18. It will be set after the class is loaded.	Breakpoints are inserted using the stop command. The first one here is a request for a breakpoint at the start of your program. The second one is a request for a breakpoint at line 18. This line must contain code. The breakpoints are not created yet.
> run run pracTest Set....... Breakpoint hit:	The run command starts the execution. It will take the program to the first breakpoint. You can ignore the screen output.
5 int[] monthTemperature = new int[12]; main[1]	This confirms that the debugger has moved forward to the first breakpoint, which is at the first line of your program code – in this case, line 5. (Whenever there is a move forward, it is to the beginning of a line.) At this stage, the prompt changes to reflect that you are running code in the main class.

(continued)

Screen display (abbreviated in places)	Explanation
`main[1] list`	The `list` command requests a numbered list of a portion of your code around the current line. (You can supply a line number to change the position for the list).
| ```
1   import java.util.Scanner;
2   class practiceTest {
3       public static void main(String[] args) {
4
5 => int[] monthTemperature = new int[12];
6   int monthNumber = 0;
7   int temperatureValue = 0;
8   for (int i = 0; i <= 11; i++)
9   {
10  monthTemperature[i] = -25;
``` | You should note that the list numbers all of the lines that you submitted to the compiler. These include continuation lines, blank lines and any comments. |
| ```
main[1] step
>
Step completed: "thread=main", practiceTest.main(),
line=8 bci=9


8       for (int i = 0; i <= 11; i++)
``` | The `next` or the `step` command can be used to move forward one line of code. The next two uses of `step` are not shown. This is the third one being used to move forward to line 8. The code has been executed each time. |
| ```
main[1] dump monthTemperature
 monthTemperature = {
0, 0, 0, 0, 0, 0, 0, 0, 0, 0, 0, 0
}
``` | The `dump` command is specific for components such as an array. The values for each element are displayed. Note that the compiler has initialised these values to zero. |
| ```
main[1] locals
Method arguments:
args = instance of java.lang.String [0] (id=465)
Local variables:
monthTemperature = instance of int[12] (id=464)
monthNumber = 0
temperatureValue = 0
``` | The `locals` command gives information about Method arguments, which you can ignore.

Then you get information about the array being used.

Finally, it shows all of the variable values. |
| ```
main[1] print temperatureValue
 temperatureValue = 0
``` | The `print` command displays the value for a specific variable (which you input). |
| ```
main[1] cont
>
Breakpoint hit: "thread=main", practiceTest.main(),
line=14 bci=51
18    if................
``` | The `cont` command takes you to the next breakpoint. |

(continued)

Screen display (abbreviated in places)	Explanation

`main[1] dump monthTemperature`
`monthTemperature = {`
`-25, -25, -25, -25, -25, -25, -25, -25, -25, -25, -25,`
`-25`
`}`
	At this point, the values you supplied for the array elements are displayed when you use the dump command.

`main[1] set temperatureValue = 15`
`temperatureValue = 15 = 15`
`main[1] set monthNumber = 6`
`monthNumber = 6 = 6`
	Because you have commented out any input or output, you need to be able to provide a value that normally would be one input.
	
	The set command can be used for this.

`main[1] next`
`Step .....`
`20    temperatureValue = temperatureValue + 1;`
`main[1] dump monthTemperature`
`monthTemperature = {`
`-25, -25, -25, -25, -25, -25, 15, -25, -25, -25, -25,`
`-25}`
	This moves to the next line you inserted so that you can see the result of you changing some values.
	
	The dump shows that the element with index 6 has been changed to 15. The debugger shows that the program works.

`main[1] exit`
 	
`C:\Users\daved\Documents\Java progs>`
	The exit command ends the debugging session and returns you to the command line prompt.

Table 11.2: Use of the Java debugging software

PRACTICE TASKS 11.2–11.3

11.2 Use the program in Practice Task 11.1 to investigate the debugger you have chosen to use. If you decide to use the built-in Java debugger, be sure to comment out all of the input and output statements.

11.3 The following pseudocode has a logical error. Write the matching Java program and run the program, inputting five values of your choosing as requested by the program. Compare the output with what you expected to get. You can use your chosen debugger to help locate the error if you wish. However, the output from the program might give you a sufficient clue as to what is wrong with the logic.

```
minimumValue ← 0
FOR i ← 1 TO 5
    INPUT nextNumber
    IF nextNumber < minimumValue
      THEN
          minimumValue ← nextNumber
    ENDIF
NEXT i
OUTPUT minimumValue
```

11.5 Using a trace table in a dry-run

In a **dry-run**, you follow the code and use the values that would be input to do the calculations defined in the code. You are in effect being the computer. You record the changing values for the variables and the output values in a **trace table**. A trace table has a column for every variable in the program plus one for the output value or values. Whenever an input statement in a program or design has a user prompt associated with it then it is helpful to include a column in the trace table to record each time the prompt is output. Each row of the table records the change following the execution of a statement in the program. A dry-run can be performed on a design recorded as pseudocode or as a flowchart or it can be performed on program code.

KEY WORDS

dry-run: a process where you follow the progress of a program by carrying out the calculations and recording the changing values for the variables.

trace table: the table used in a dry-run to store the changing values for the variables, outputs and user prompts.

DEMO TASK 11.1

You have decided to create a program to find the result of an integer division. Specifically, the program should find the number of times that a number y can divide into a number x.

Solution

Because you are wishing to develop your programming skills, you have decided not to use the operator normally used for integer division. Instead, you produce the following pseudocode design for the program.

```
numberOfTimes ← 0
INPUT x
INPUT y
WHILE x > y DO
    x ← x - y
    numberOfTimes ← numberOfTimes + 1
ENDWHILE
OUTPUT numberOfTimes
```

You decide to check the logic of the design by carrying out a dry-run. For this dry-run, you decide to use 50 for the x input and 15 for the y input. Table 11.3 shows the trace table for the dry-run. You have only entered a value in a column when it has changed from the previous row. You have decided to include comments in the table but you realise that these would not normally be expected.

x	y	numberOfTimes	Output	Comments
		0		Initialisation.
50				The x value is input.
	15			The y value is input.
35		1		x is reduced by 15, numberOfTimes is incremented by 1. ENDWHILE returns to the WHILE condition check. As x > y, the loop continues to run.

(continued)

CONTINUED

x	y	numberOfTimes	Output	Comments
20		2		x is reduced by 15, `numberOfTimes` is incremented by 1. `ENDWHILE` returns to the `WHILE` condition check. As x > y, the loop continues to run.
5		3		x is reduced by 15, `numberOfTimes` is incremented by 1. `ENDWHILE` returns to the `WHILE` condition check. As x < y, the loop exits.
			3	The value in `numberOfTimes` is output.

Table 11.3: Trace table for the pseudocode design

The result you obtained was the one you expected, so you believe that the logic is fine.

PRACTICE TASK 11.4

On reflection, you decide that just one dry-run is insufficient to check the logic. Repeat the dry-run in Demo Task 11.1, but this time use the value 60 for x and 15 for y.

When you have completed the dry-run, can you see what the logical error is in the design?

CHALLENGE TASK 11.1

Flowchart 11.1 shows a flowchart design for a program that is to calculate the number of tins of paint needed when painting a wall. The design includes calculations that involve the dimensions of the wall, the area covered by one tin and the number of coats that will be needed. All values are stored as integers. The calculation includes an integer division, which gives an integer answer.

1 Carry out a dry-run of this design using the values:

- length 6 m

- width 5 m

- coverage for one tin 7 m^2

- number of coats 2.

CONTINUED

2   Carry out a dry-run of this design using the values:

- length 3 m

- width 2 m

- coverage for one tin 7 m²

- number of coats 1.

3   Can you see how many logical errors there are in the flowchart design?

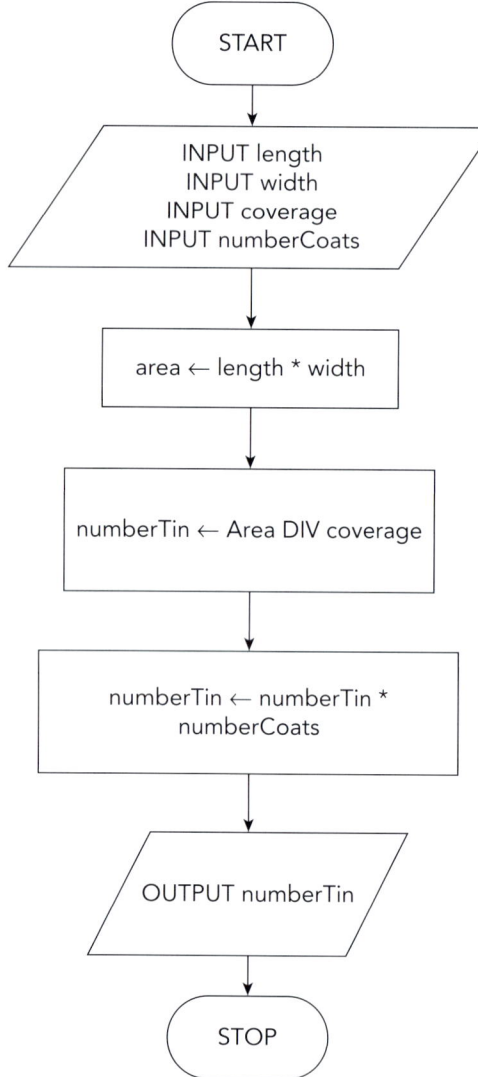

Flowchart 11.1: A flowchart for calculating the number of tins of paint required

SUMMARY

A syntax error is where the programming language has been used incorrectly. The compiler identifies syntax errors.
A runtime error is where the combination of the data values input and the processing of these values causes the program to halt with an error message.
A logical error is one that allows a program to run to completion but where the resulting output is incorrect.
Debugging is the process of locating and correcting errors. A debugger is software that assists in debugging.
A debugger allows step-by-step execution of a program.
A debugger allows breakpoints to be inserted to move the execution forward and then temporarily halt the program.
A debugger will allow variable values to be examined and possibly altered.
A dry-run can be used to locate errors. It requires the programmer to do the calculations and record the changing values for variables in a trace table.

END-OF-CHAPTER QUESTIONS

1   Read the following three scenarios and identify which is the *logical error*, which is the *syntax error* and which is the *runtime error*.

   a   A programmer has completed the first subsystem of a new social media app; however, it fails to compile. The IDE highlights a line of the code, stating that there is an unexpected end of line.

   b   During beta testing, one of the testers reported that the new music creation application he was testing crashed with the following error message: 'ZeroDivisionError: division by zero'.

   c   Another beta tester of the same music creation application reported that when she selected piano, the application kept sounding like a harpsichord.

2   a   State the four types of test data.

   b   For each type, give a brief explanation.

3   Explain three features that would be provided in a debugger.

> Chapter 12
Programming scenario task

IN THIS CHAPTER YOU WILL:

- learn about how you might approach providing an answer to a programming scenario question in the Algorithms, Programming and Logic Paper

- practise using techniques for program design and coding learnt in earlier chapters

- use the Input, Process, Output framework as a basis for organising your solution.

The information in this section is based on Cambridge IGCSE, IGCSE (9–1) and O Level Computer Science syllabuses (0478/0984/2210) for examination from 2023. You should always refer to the appropriate syllabus document for the year of your examination to confirm the details and for more information. The syllabus document is available on the Cambridge International website at www.cambridgeinternational.org.

Introduction

This chapter supports the 'programming scenario task' in the new syllabuses. The chapter will take you through a demo task in the style of a programming scenario task. It will provide step-by-step suggestions as to how you might break the task down and construct an appropriate solution. This type of task will require you to provide a coded solution to a problem outlined in a scenario. You can use pseudocode or programming code for the solution. The coded solution required will be more extensive than what has been required in most of the tasks provided for you in earlier chapters of this book. If there are areas that you don't understand, it might help to re-visit the relevant section of the book.

12.1 Suggestions for tackling the question

There are three constraints to consider.

1   You will be writing your answer to the scenario question in the examination paper, as you will for all of the questions in the paper. You will not have any software available for editing or for cutting and pasting. It will be important to write your answers with the content initially spaced out to allow for later additions or details to be inserted in the spaces you have left.

2   You will not have your usual software for running and testing program code. Because of this you do not have to worry about minor syntax errors. You are not expected to provide a perfect solution.

3   There is a limited time that you have for creating your solution. If this were a coursework exercise with no strict time limit, you could be expected to provide:

   a   a structure chart design

   b   a flowchart design

   c   a pseudocode design

   d   the program code

   e   validation of input

   f   a test plan.

Instead you only have to present pseudocode or program code. The scenario could possibly require validation of input, but this would be made clear in the question.

Although no formal preliminary design techniques will be expected, it is important that you organise your thinking about the problem. It will be helpful to start with the idea that the solution will have three components:

- data input
- data processing
- data output.

Based on this, the following checklist is a suggestion for the questions you might ask yourself and the order in which you might consider them.

1   What output is to be achieved?
2   What data is already provided?
3   Are any variables, constants or arrays required that have not been provided?
4   Therefore, is any input needed?
5   Will selection constructs be needed as part of the processing?
6   Will iteration constructs be needed as part of the processing?

You might wish to highlight any such information included in the description of the scenario requirement given. It will certainly be helpful if you write down answers to these checklist questions. However, be brief so you do not spend too much time on this.

When you come to writing your code, you should aim to:

- use appropriate variables and data structures for the storage of data
- use meaningful names for these variables and data structures
- use appropriate programming techniques
- use a logical structure for the code
- include the specific content that has been requested in the scenario
- include comments to explain the actions performed by the code.

If you are short of time, you can concentrate your efforts on those parts of the solution where you are most confident that you know what is needed.

DEMO TASK 12.1

The problem scenario concerns an automatic washing machine which contains a small embedded computer system. This system has a program installed that controls the operation of the machine. You have been asked to write the code for the part of this program that handles the interaction with the user of the washing machine.

CONTINUED

The user has to identify the type of load for the wash. Examples of types of load are White cotton, Cotton colours and Wool. The type of load has to be chosen so that appropriate controls can be set for the wash. Table 12.1 shows the five preset features that the washing machine controls together with the range of possible values that can be chosen.

Features to be controlled	Minimum value	Maximum value
Wash time (in minutes)	10	70
Agitation speed (number per second)	01	10
Wash temperature (in °C)	20	90
Rinse time (in minutes)	05	30
Rinse speed (rotations per second)	01	10

Table 12.1: Values of the features of a washing machine

The program running the machine has access to a number of text files. Each text file contains a list of values for one type of load. The name of each text file matches the name of a particular type of load. The line of text in each text file has five two-digit numbers separated by single spaces.

For example, there will be a text file named `wools.txt` *for the wool load containing:*

```
30 02 30 30 02
```

The requirements for the part of the program that you are asked to write are as follows:

- *The name for the type of load is input.*

- *The control data for this load is read from the appropriate file.*

- *The data from the file is split into five separate values.*

- *The data for the wash temperature is checked to ensure that it is within the allowed range.*

- *If the temperature is not in the allowed range, the program sends an error message and stops.*

- *The total time for the wash is calculated and output in hours and minutes.*

- *The wash temperature is output.*

- *The text file is closed.*

Solution

We are now ready to consider how you might proceed to produce a solution to the problem scenario. As this is an extended task, the rest of this chapter describes how to go about completing a full solution to this scenario task.

12.2 Coding a solution

You have the option to provide pseudocode or programming code. Let us assume that you choose to write Java code. You have not been asked to write a complete program; you just need some lines of code to match the specific requirement.

If you ask the questions suggested in the checklist given in Section 12.2, you might write down answers like the following.

1   Total time and temperature, unless there is a problem with temperature.
2   Data in files assumed.
3   Yes – string for reading the text file then variables for control values.
    No arrays or constants.
4   The load type then read the file.
5   One for checking temperature.
6   No iterations.

At this stage you could decide to set out a list of comments, using plenty of space to allow code to be inserted between them. These could be:

```
// Input the load type
// Open and read the text file
// Split the file data into five values
// Check the temperature
// If temperature reasonable, carry out calculations
// Output values
// Close the file
```

Presenting these as comments gives you a framework for writing your code but can also lead to marks being awarded.

The following is an example of what would be a good solution. You should note that at this stage, you would expect to leave some comments unchanged but for others a rewording or additional comment might be needed.

```
// Input the load type

System.out.println("Enter the name of the type of load");
Scanner myObj1 = new Scanner(System.in);
String progName = myObj1.nextLine();

// Open and read the text file

String inputfilename = progName + ".txt";
FileReader inputObj = new FileReader(inputfilename);
Scanner InObj = new Scanner(inputObj);
String strInput = InObj.nextLine();

// Split the file data into five values
// Using substrings creates string variables which have to
// be converted to integers

String first = strInput.substring(0,2);
int washTime = Integer.parseInt(first);
```

```
String second = strInput.substring(3,5);
int agitationSpeed = Integer.parseInt(second);
String third = strInput.substring(6,8);
int washTemp = Integer.parseInt(third);
String fourth = strInput.substring(9,11);
int rinseTime = Integer.parseInt(fourth);
String fifth = strInput.substring(12,14);
int rinseSpeed = Integer.parseInt(fifth);

if ( washTemp < 20.0 || washTemp > 90.0)
System.out.println("Program error");
// Check the temperature
else
{
// If temperature reasonable, carry out calculations

int minutes = washTime + rinseTime;
String hours = String.valueOf(minutes/60);
String mins = String.valueOf(minutes%60);
String duration = hours + " hours " + mins + " minutes";

// Output values

System.out.println(duration);
System.out.println("Wash temperature: " + washTemp + "
    degrees C");

}

// close the file

inputObj.close();
```

You do not have to write the code by starting at the beginning and working through to the end. You can write those parts first with which you are most confident. This will be particularly sensible if you are short of time for answering the scenario question.

12.3 Final thoughts

Once you have finished what you can, it is useful to go through a mental checklist to try to ensure you get as many of the marks available as possible. Table 12.2 shows a suggested checklist.

Code feature	Questions to ask
Inputs:	Has any data not already provided been input?
Data storage:	Have I used the variable names provided? Are my own variable names meaningful? Are my variables all of the correct data type? Are the data structures I have used in my algorithms the best ones to use? Do my data structures store all the data they should?
Processes:	Have I used appropriate program constructs? Is the logic of the coding correct?
Outputs:	Does my program produce all the outputs required? Are the outputs in the form asked for?
Other:	Is my solution sensibly commented?

Table 12.2: Mental checklist

If you have worked through the first 11 chapters of this book, you already have all the programming techniques necessary to answer these types of questions. The best way to prepare now is to gain some practice.

Practise solving tasks, allowing yourself a limited time for completion. You might begin by allowing longer than 30 minutes before trying to produce an answer in the time limit suggested for the scenario task question. The following five end-of-chapter questions have been provided specifically for this chapter. You could also attempt solutions for tasks provided in earlier chapters.

In the exam, you are allowed to provide a pseudocode or a programming solution. You may wish to try both options. However, the solutions chapter to this book only includes Java code solutions. Whichever you choose, you are expected to provide comments to explain the code.

SUMMARY

Identify the inputs, processes and outputs of the scenario.
Use appropriate messages when inputting and outputting data.
Create a structured plan using commenting.
Use meaningful identifiers for variables, constants and subroutines.
Select appropriate data types if not provided.
Choose efficient algorithms for processing data.
Ensure outputs are in the form requested.
Add further comments to explain your code, if necessary.

END-OF-CHAPTER QUESTIONS

1 A program is needed to provide a guessing game for two players.

Player 1 is invited to play.

Player 1 is asked to guess a number between 1 and 10. The input value is compared to a randomly generated number. If the numbers match, the score for Player 1 is incremented.

Whether the numbers match or not, the randomly generated number is stored if it is the maximum so far for Player 1. The randomly generated number is also added to a running total for Player 1. The process is repeated for Player 1 another nine times.

Player 2 is invited to play.

There is a repeat of the guessing of ten numbers by Player 2.

The scores for the two players are compared. The winner is the player with the highest score.

If the scores are equal, the maximum random numbers are compared. If one player has a higher maximum value, that player is the winner.

If the maximum random number values are equal, the running totals are compared to find the player with the highest value. That player is the winner.

If the running totals are the same, the game is drawn.

The program should output either the identity of the winner or a statement that the game is drawn if no winner.

2 A program is needed to help identify how healthy certain food items are. The program will calculate health ratings using data stored in two arrays.

The following tables show examples of some of the data stored:

`snackName` array

Index	0	1	2	3	4
Name	Caramello	Oaty	Chocbar	Slimaid	Nutter

`nutrition` array

Index	0	1	2	3	4
Carbohydrate /100 g:	69.3	65.1	59.8	35.4	54.4
Sugars /100 g:	59.9	32.3	58.7	27.1	44.8
Protein /100 g:	4.4	5.1	5.8	3.1	9.9
Salt /100 g:	0.42	0.48	0.2	0.32	0.45

The 1D array `snackName` contains the names of popular sweetshop snacks. The 2D array `nutrition` contains the mass (in grams per 100 g) of carbohydrates, sugars, protein and salt for each snack. You should note that the sugars are a subset of the carbohydrates. The arrays have been designed so that the index values identifying the columns in the `nutrition` array match the index values in the `snackName` array.

The rules used by the program are as follows:

a Snacks are awarded a health index using the formula:

The health index = Carbohydrate + (Sugars * 2) + (Salt * 100).

b There are three ratings based on the value of the health index:

- Red = for when the health index is 200 or above.
- Orange = for when the health index is less than 200 but at least 150.
- Green = for when the health index is less than 150.

CONTINUED

The program code for creating the arrays described above and storing the values has already been written. In addition, a variable `sampleSize` has been created, which has a value indicating the number of snacks for which data is stored in the program.

You need to provide the remaining code for the program.

This program code must do the following:

- Calculate the health index for each of the snacks.
- Calculate the average value of the health indexes of all of the snacks, rounded to the nearest whole number.
- Output, for each snack:
  - name
  - percentage of snack that is carbohydrates, rounded to the nearest whole number
  - the health index
  - the health rating (Red, Orange or Green).
- Output the average rounded value of the health indexes of all of the snacks.
- Calculate, store and output the number of Red, Orange and Green snacks in the whole sample for which data is stored.

3 A bank is developing a system for handling interactions with customers. The system will contain a suite of individual programs. You are requested to provide a program for checking the customerID when a customer wishes to undertake an online transaction. The program requirement is as follows:

- The program will begin by requesting input of the customerID from the customer.
- A correct customerID consists of four lower case letters followed by two numeric digits.
- The program will check that the input matches this required format.
- This will require three checks:
  - that the length of the input string is six
  - that the first four characters are lower case letters
  - that the last two characters are numeric digits.
- If a check finds an error, the program must output a message indicating the type of error that has been found.
- Following this error message, the customer is offered the chance to input the whole customerID again.
- The customer will only be allowed three attempts.
- If three incorrect attempts have been entered, the program will output a message to state that no further attempts will be allowed.
- If a customer has entered a customerID in the correct format, the program concludes by outputting a message to say that the input has been accepted.

4 Thirty schools are taking part in a weather monitoring project. The project lasts for seven days. Each school records both the minimum and the maximum temperature (in degrees Celsius) at their location for each of the seven days.

You are asked to provide a program that meets the following requirements:

- The program calculates the daily temperature range (maximum temperature minus minimum temperature) for each of the schools.
- The program calculates the average daily temperature range over the seven-day period for each school.
- The program calculates the average maximum temperature over the seven-day period for each school.

CONTINUED

- The program outputs, for each school:
  - the name of the school
  - the average temperature range
  - the average maximum temperature.
- The program also calculates, stores and outputs the maximum temperature and the minimum temperature recorded at any of the schools.

You have already started to create the program. You have written the declaration statements and the input statements for recording the data for the following three arrays:

1 A 1D array named `school` to store the names of the schools.

2 A 2D array named `maxTemp` to store the maximum temperature recorded at each school for each of the seven days.

3 A 2D array named `minTemp` to store the minimum temperature recorded at each school for each of the seven days.

Each 2D array has 30 rows with indexes matching those for the 1D `school` array.

You are now required to write the remaining code for the program.

5 A running club has a competition where all members take part. The distance each member can run in 30 minutes is recorded. Each member is allowed up to three attempts.

Members are allocated to categories based on the distance achieved in their best run. The criteria are shown in the following table.

Category	Distance achieved in best run
Elite	Greater than or equal to 7 kilometres.
Championship	Greater than or equal to 5 kilometres and less than 7 kilometres.
Club	Less than 5 kilometres.

A program is needed that meets the following requirements:

- The program must calculate the average distance run for each member. Where a member has completed fewer than three runs, the average will be based only on the number of runs they completed.
- The program must calculate the best distance for each member.
- The program must output, for each member:
  - name
  - number of runs completed
  - average distance covered
  - best distance
  - category awarded.

You have already started the program. You have declared and assigned a value for a variable `numberMembers`. You have written the declaration statements and the input statements for recording the data for the following two arrays:

1 A 1D array named `members` to store the names of the members.

2 A 2D array named `distance` to store the values for each run for each member.

The index values for the rows in the array `distance` match the index values in the array `members`.

You are now required to write the remaining code for the program.

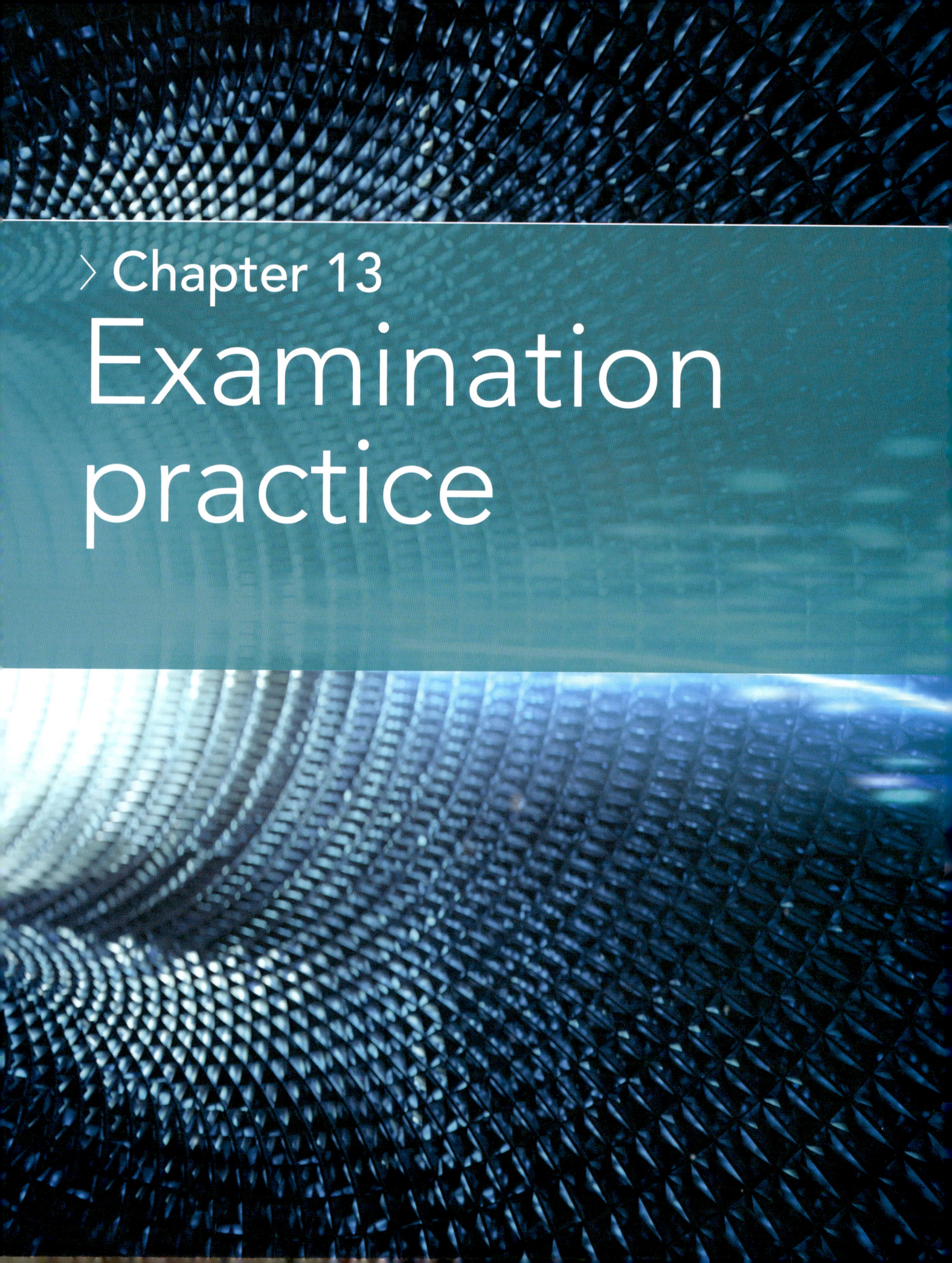

> Chapter 13
Examination practice

Exam-style questions and mark scheme guidance have been written by the authors. In examinations, the way marks are awarded may be different. References to assessment and/or assessment preparation are the publisher's interpretation of the syllabus requirements and may not fully reflect the approach of Cambridge Assessment International Education.

Introduction

This chapter includes a series of exam-style questions. This chapter aims to bring together all the skills you have developed and all the knowledge you have gained throughout this book. The questions will test your understanding of key programming concepts you have learned and offer you an opportunity to identify any gaps in your understanding. If you are unsure about how to answer a particular question, it may help to re-visit the relevant section of the book. There are full, suggested solutions in the solutions chapter in the digital part of this resource. Remember that there is often more than one solution. Good luck!

Some points to note:

- When a question content includes pseudocode, the pseudocode is presented in the same way that it would be presented in a question in a real examination paper. The pseudocode follows the rules defined in the syllabuses. In particular, all identifiers begin with an upper case letter.

- In presenting an answer you can continue to write pseudocode in the same style that you have been using when providing solutions to tasks in earlier chapters.

- The overriding requirement is for the logic of your pseudocode to be correct.

- However, if a question includes identifiers you must ensure that these are used in your pseudocode identically as provided in the question.

- Finally, once you have defined an identifier you must continue to use it without change throughout your code.

EXAM-STYLE QUESTIONS

1   A school is holding a vote to decide on the name for a new science block. The two options are Faraday and Curie. Each of the 601 students will vote. An algorithm is to be created for an electronic voting system that will record the vote of each student. You can assume that all the students will make a valid vote. The system will output the most popular name and the percentage of students that voted for that name.

Write an algorithm for the voting system, using pseudocode or a flowchart.

[6]

2   The following pseudocode inputs the scores achieved by students in a test. The maximum score in the test was 200. A value of –1 stops the input. The algorithm outputs the highest, lowest and average score.

```
DECLARE Marks: INTEGER
Marks ← 0
DECLARE High: INTEGER
High ← 0
```

CONTINUED

```
DECLARE Low: INTEGER
Low ← 0
DECLARE Total: INTEGER
Total ← 0
DECLARE Count: INTEGER
Count ← 0
DECLARE Average: REAL
Average ← 0.0
OUTPUT "Enter Student Marks"
INPUT Marks
WHILE Marks <> -1 DO
    Total ← Total + Marks
    Count ← Count + 1
    If Marks <= High
      THEN
         High ← Marks
      ELSE
        IF Marks < Low
           THEN
              Marks ← Low
          ENDIF
     ENDIF
     OUTPUT "Enter Student Marks"
     INPUT Marks
ENDWHILE
Average ← Count / Total
OUTPUT High, Low, Average
```

There are four errors in the algorithm.

Locate those errors and show how the error could be corrected.

[8]

3   A system accepts 13-digit ISBNs. The ISBN includes a check digit.

a   Explain how a check digit is used to verify the input.

[4]

b   Each of the following pseudocode fragments represents a validation.
    For each example in the table, state the type of validation that is being completed.

Pseudocode	Type of validation

`INPUT ISBN`
`IF ISBN = ""`
`  THEN`
`    OUTPUT "Error"`
`ENDIF`	

`INPUT ISBN`
`IF LENGTH(ISBN) < 13`
`  THEN`
`    OUTPUT "Error"`
`ENDIF`	

[2]

CONTINUED

c   This pseudocode has been written to validate user input for a system that accepts only integer values.

```
DECLARE UserValue : INTEGER
INPUT UserValue
WHILE UserValue <=10 OR UserValue >= 200 DO
    OUTPUT "Out of range, re-enter value"
    INPUT UserValue
ENDWHILE
```

The code is to be tested using normal, abnormal and extreme data.

Give one example of each type of test data (normal, abnormal and extreme) that could be used to test the validation algorithm.

[3]
[Total: 9]

4   The algorithm inputs a series of integer values. A negative value stops the input.

```
DECLARE Number : INTEGER
Number ← 0
DECLARE Count : INTEGER
Count ← 0
DECLARE Total : INTEGER
Total ← 0
DECLARE Large : INTEGER
Large ← 0
REPEAT
    INPUT Number
    Count ← Count + 1
    IF Number >= 0.0
      THEN
        Total ← Total + Number
        IF Total MOD 20 = 0
          THEN
            Large = Total DIV 20
        ENDIF
    ENDIF
UNTIL Number < 0.0
Count ← Count -1
OUTPUT Large * Count
```

Create and complete the trace table for the following sequence of inputs:

15, 0, 14, 0, 12, 19, 11, −6.

Use the following headings in your trace table.

Number	Count	Total	Large	Output

[5]

CONTINUED

5 An array `netName` holds the network usernames of 600 employees. The array has not been sorted.

An algorithm is required that will search the array for a specific username. If the username is in the array, the algorithm will output the index location of the array that holds the specific username. If the username is not in the array, the algorithm will output the message 'No Record'.

a Write the search algorithm, using pseudocode or a flowchart.

[5]

The array `netName` has been sorted in alphabetical ascending order.

b Explain how your algorithm could be made more efficient now the array `netName` is sorted.

[3]

[Total: 8]

6 A 2D array is used in an electronic board game. The pseudocode to declare the array is:

```
DECLARE Board : ARRAY[0:8, 0:8] OF STRING
```

When each game starts, each index location in the array should be assigned the value 'Free'.

Use pseudocode to write an algorithm to assign the value 'Free' to all index locations in the array `Board`.

[3]

7 A business specifies that product numbers meet these requirements:

- The product number must begin with the letters 'PROD'.
- The product number must be 10 characters long.

The product numbers PROD1556DT and PRODa45629 are valid product numbers.

A system is required that will check that product numbers meet the requirements. The system will output the message 'Accepted' or 'Rejected', depending on whether the product number input meets the rules.

Write the algorithm that will check that product numbers meet the requirements.
Use pseudocode or a flowchart.

[6]

8 Rewrite the following pseudocode using a CASE statement.

```
DECLARE Score: INTEGER
Score ← 0
INPUT Score
IF Score > 100
  THEN
    OUTPUT "Excelling"
  ELSE
    IF Score > 80
      THEN
        OUTPUT "Good"
    ENDIF
  ELSE
    IF Score > 60
      THEN
        OUTPUT "Acceptable"
    ENDIF
  ELSE
    OUTPUT "Below expectations"
ENDIF
```

[4]

CONTINUED

9 A function `convert` is required which will take as a parameter the temperature in degrees Celsius and output the equivalent temperature in degrees Fahrenheit.

The formula to convert from Celsius (C) to Fahrenheit (F) is F = C * 1.8 + 32.

Write the pseudocode for the function `convert`.

[3]

10 Programmers make use of variables and constants.

a Explain **one** difference between a variable and a constant.

[2]

Variables and constants are declared using data types.

b Provide the most appropriate data types for each variable:
- `students` – holds the number of students in a school.
- `id` – holds ID numbers. An example is MR145T.
- `weight` – holds the weight of parcels in kilograms.
- `passed` – records whether a person has passed a test.
- `name` – holds the name of students at a school.
- `grade` – holds an examination grade A, B, C, D or E.

[6]
[Total: 8]

11 A tennis club classifies its members as:
- child if aged less than 11
- junior if aged 11 to 17
- adult if aged 18 to 59
- senior if aged 60 or older.

The following is a partially completed flowchart design for a program which would be used to store the classification of each member in a variable called classification.

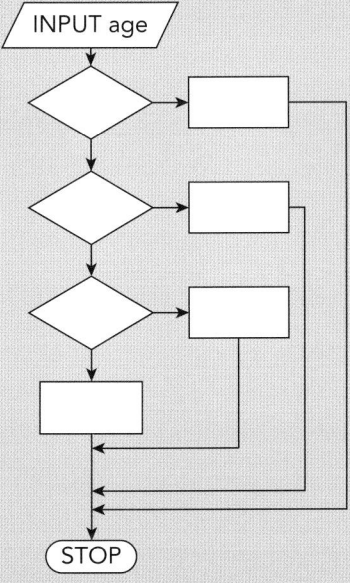

Copy and complete the flowchart to show the design of an appropriate algorithm for this task.

[8]

CONTINUED

12 a Define the term **variable**.

[2]

b A school stores data in a computer system regarding students' performance in tests. The following table shows some example data that would be stored by the system. Note that the column headings in the table are definitions of what the values represent. They are not variable names.

Student name	Student identifier	Mark	Percentage	Comments	Re-test needed
A Patel	346	60	80.5	improved	No
W Chi	231	65	87	steady	No
E Brown	458	33	44	not your best	Yes
M Santana	023	42	56.5	better	No

Identify the data type you would use for each data element.

Data element	Most appropriate data type
Student name	
Student identifier	
Mark	
Percentage	
Comments	
Re-test needed	

[4]

[Total: 6]

13 a i State three reasons why a programmer would choose to use a subroutine.

[3]

ii State two differences between a function and a procedure.

[2]

b A subroutine is to be used to convert a distance in miles into a distance in kilometres. The calculated value is stored in a variable with the identifier KmDistance.

i Choose a suitable name for the subroutine.

ii Write a pseudocode statement showing how the subroutine would be used if it had been written as a procedure.

iii Write a pseudocode statement showing how the subroutine would be used if it had been written as a function.

[3]

[Total: 8]

CONTINUED

14 A program contains a variable `oldString` which holds the value 'someWORD'. Using pseudocode or program code, write string-manipulation statements that will assign values to another variable called `newString` from the value currently in the variable `oldString`.

   a   `newString` is to hold the value 'SOMEWORD'.

   b   `newString` is to hold the value 'D'.

   c   `newString` is to hold the value 's'.

   d   `newString` is to hold the value 'WORD'.

[4]

15 A club is developing a system for dealing with applications for membership.
When people apply to join the club, they are asked for personal details and the name of an existing club member who will recommend them. The club will then obtain supporting details from the existing club member. The decision as to whether the application has been accepted is stored. The applicant will be sent either a 'Welcome to membership' or a 'Sorry application rejected' letter.

The following structure diagram has been created as a design for the system. You need to provide labels for each box in the diagram.

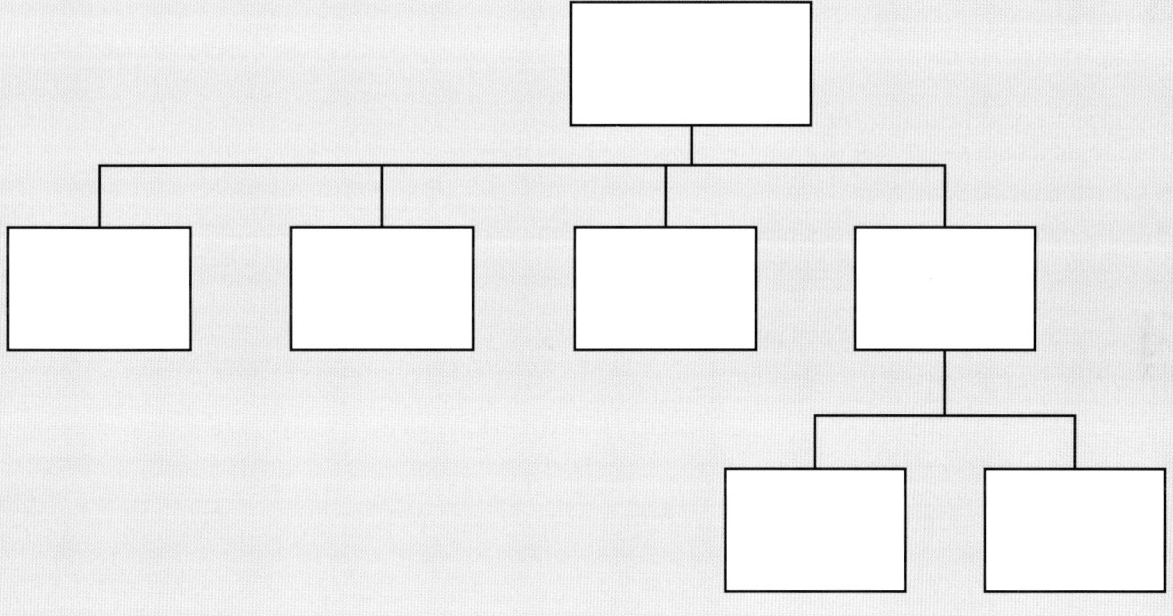

[5]

16 Consider the following fragment of pseudocode:

```
FOR i ← 1 TO n - 1
    Swapped ← false
    FOR j ← 0 TO n - 1 - i
      IF List[j] > List[j + 1]
        THEN
            Swap(list[j], list[j+1])
            Swapped ← TRUE
      ENDIF
    NEXT j
    IF Swapped = false
      THEN
          break
NEXT i
```

   a   Identify the purpose of this algorithm.

[1]

   b   Identify and explain the programming constructs that are used.

[5]
[Total: 6]

17 A student writes a program for a running club.
The program requires the user to input the following information:

- First name and last name.

- The time it took them to complete each of their last five marathons.
  Each time is rounded to the nearest minute.

The program will output the average time for the five marathon runs.
The average is output in hours and minutes.

This is the pseudocode for the algorithm:

```
FUNCTION f(i: INTEGER, j: INTEGER, k: INTEGER, l
  INTEGER, m: INTEGER) RETURNS INTEGER
    t ← i + j + k + l + m
    RETURN ROUND(t/5)
ENDFUNCTION
INPUT Name1
INPUT Name2
FOR i ← 1 TO 5
    CASE OF i
      1: INPUT a
      2: INPUT b
      3: INPUT c
      4: INPUT d
      5: INPUT e
    ENDCASE
NEXT i
Ave ← f(a,b,c,d,e)
OUTPUT Name1 , " " , Name2 , " average marathon time:"
OUTPUT (Ave DIV 60) , " hrs" , (Ave MOD 60) , " mins"
```

CONTINUED

a   Rewrite the algorithm, keeping its logical flow, but make it easier to read and more manageable.

[6]

b   Another programmer tells the student that the algorithm is overly complex and could be produced without the use of the FUNCTION or the CASE statement. Rewrite the algorithm without using FUNCTION or CASE.

[6]

[Total: 12]

18 A simple game app keeps track of the number of lives a player has in a variable called `Lives`. Players can either gain or lose lives, depending on the event that happens; one of the two procedures in the following pseudocode snippet are called.

```
Lives ← 5
GameOver ← FALSE
PROCEDURE LoseLife()
    MinLives ← 0
    IF Lives > MinLives
      THEN
         Lives ← Lives - 1
    ENDIF
    IF Lives = 0
      THEN
         GameOver ← TRUE
    ENDIF
ENDPROCEDURE

PROCEDURE GainLife()
    MaxLives ← 5
    IF Lives < MaxLives
      THEN
         Lives ← Lives + 1
    ENDIF
ENDPROCEDURE
```

a   Give the identifiers of all the local variables.

[1]

b   Give the identifiers of all the global variables.

[1]

c   Explain the difference between a local variable and a global variable.

[2]

d   Explain why it is not a good idea to make all variables global.

[2]

[Total: 6]

CONTINUED

19 A program needs to read a value from the text file `Score.txt`, increase it by 1 and then store the new value back in the text file. The program must also check that the new score has not reached 10. If the score is 10 or above then the program calls the `gameOver()` procedure.

Write an algorithm, using pseudocode or program code, to perform this part of the larger program.

[6]

20 A program is required that creates usernames and passwords from users' first and last names. First the program must ask a user to input their `firstname` and then their `lastname`. It then creates a username from the first two letters of the `firstname` and the last five letters of the `lastname`. The `username` is entirely lower case. The program then creates a `password` that is made by creating a random six-figure number that does not start with 0. Finally it outputs the `username` and `password` for this user.

Assumptions:

- Ignore possible duplicate usernames.
- Users' first and last names only contain letters (no hyphens or apostrophes or accents).
- Users' first and last names may contain a mixture of upper case and lower case letters.
- Users' first and last names always contain more than 3 letters and 5 letters respectively.
- For the purpose of this question, there is no storage of the username and password produced.

Write an algorithm, using program code, to perform this part of the larger program.

[6]

21 The pseudocode design for simple computer game currently stores a map of rooms in a 10 × 10 grid in 10 arrays declared as follows:

```
DECLARE Row0 : ARRAY[0:9] OF STRING
DECLARE Row1: ARRAY[0:9] OF STRING
DECLARE Row2: ARRAY[0:9] OF STRING
etc.
```

a Why would it be better to store the map data in a single 2D array?

[3]

b Provide the declaration pseudocode for a 2D array, `Map`, that can be used to store all the information in the ten 1D arrays currently used.

[1]

c A reference to an image is stored in `Map[2,2]`, e.g. `'hall1.png'`. Stored in `Map[4,4]` is a reference to another image. During the game, these two image references need to be swapped around. Write the pseudocode that would swap the two image references in the map array. In your answer, you must assume that the current images being shown in the 'rooms' have to be found, by your program, from the array.

[3]

[Total: 7]

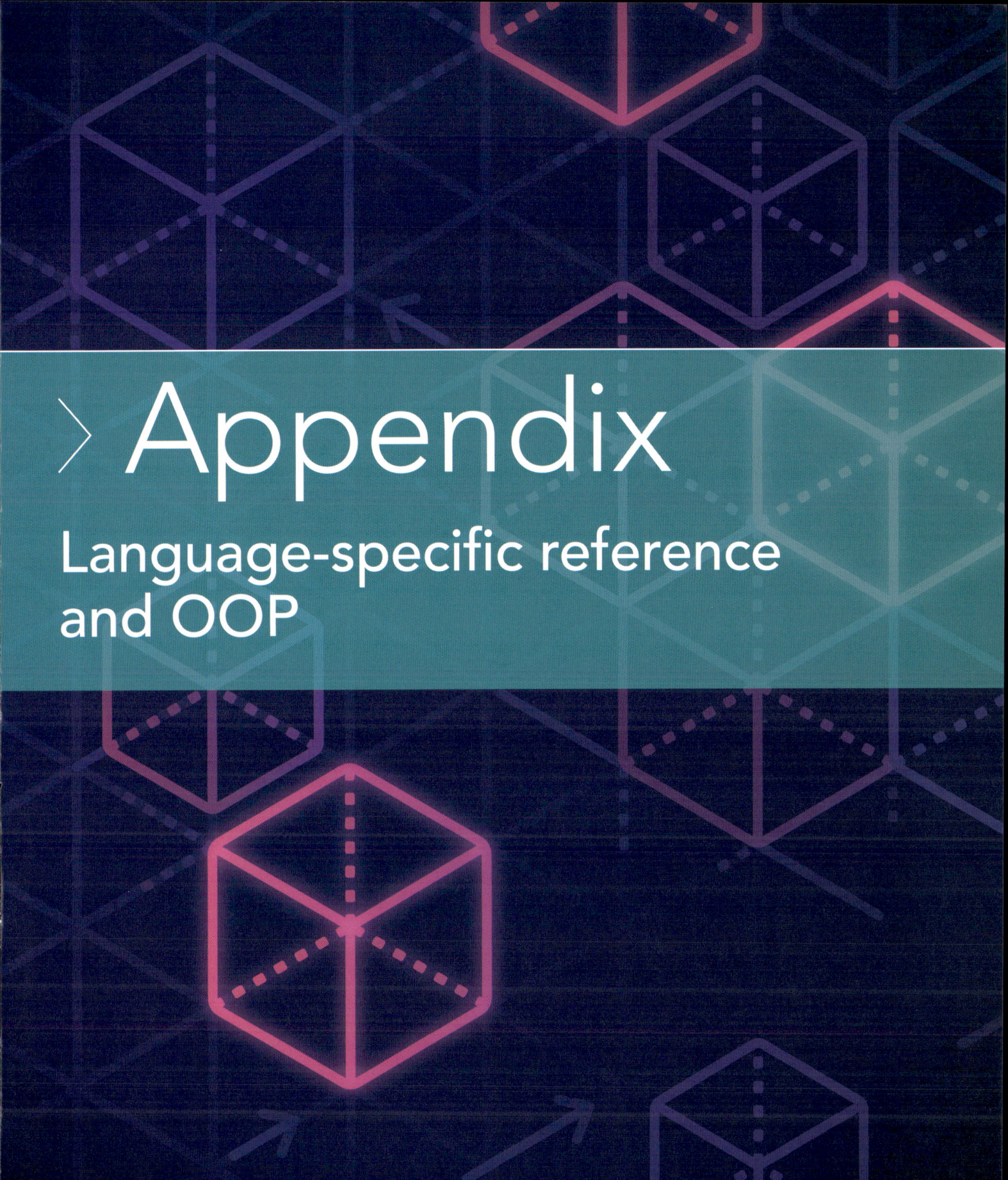

› Appendix
Language-specific reference and OOP

Key words

The following is a list of key words which are reserved in Java and must not be used as an identifier. Most of these key words have been unchanged since the first ever version of Java. Occasionally, new ones are added when a new version of Java is developed. The list here is complete at the time of writing.

abstract	assert	boolean
break	byte	case
catch	char	class
const	continue	default
do	double	else
enum	extends	final
finally	float	for
goto	if	implements
import	instanceof	int
interface	long	native
new	package	private
protected	public	record
return	short	static
strictfp	super	switch
synchronized	this	throw
throws	transient	try
void	volatile	while
var	yield	

In addition, you should not use `true`, `false` or `null` as an identifier.

Object-oriented-programming (OOP)

The focus here is to explain those aspects of OOP that you meet when writing simple programs of the type introduced in this book. There is a wide range of concepts fundamental to OOP which we do not need to consider.

It seems obvious that the fundamental concept in OOP must be the object. However, there is a strong argument for saying that it is in fact the class which is fundamental. The reason for saying this is that you cannot create an object unless you have already created a class. By contrast, you can create and use a class without ever creating a related object.

A class is a data type that is far from simple. A class can store program code and data in the same way that an object can. In particular, the program code that you write is created as part of the definition of a class.

It will now be useful to consider the framework that you have been using for running your programs.

```
class MyClass {
    public static void main(String[] args) {
    ****
}}
```

Here you are defining your own class with the name `MyClass`.

There is no data stored in this class. There is just one program. In OOP, programs defined in a class or an object are called methods. In this case you are creating a method with the name `main`. It is a requirement in Java that the first method to be used must be named `main`. There must be only one class defined which has a method with this name. There is then a further requirement that the name of the class that includes the `main` method must be used in the filename that contains the code. In this case, the filename must be `MyClass.java`. This is, of course, case sensitive.

The following are some comments about the line of code that defines the main method:

- `public` is called a modifier and indicates that the method can be used by any other class or object. You must include this when defining the `main` method. It can be noted that the class definition could also use this:

  ```
  public class MyClass
  ```

  but it is not essential in this simple framework for creating a program.

- `static` is a keyword that allows a method to be used without the need to create an object from the class.

- `void` is a keyword that states that the method does not return a value. In other words, the method is working as a procedure. If a method were required to work as a function, the keyword `void` would be replaced by a data type for the returned value.

- `(String[] args)` is a definition of the arguments used by the method. It states that `args` is the name of a `String` array. You can in fact use anything you like for the name of the array but there is no benefit in changing it. The array stores values of any arguments you supply when using the Java command. For example, if you started the program by typing in the following at the command line prompt:

  ```
  java MyClass 35 name
  ```

  then you would have `args[0]` containing "35" and `args[1]` containing "name". Your program could then use these wherever needed.

Before looking at other aspects of OOP in Java, it will be helpful to introduce some terminology that you may encounter in your studies.

An object can be said to have an **identity**, a **state** and a **behaviour**. The **identity** is the name given to the object. The **state** is defined by the values stored in the object. These are called fields, attributes or variables. The **behaviour** refers to the methods which operate as functions or procedures. The state of the object can be changed by the use of a method. The methods and the fields are together called the members.

When an object is created, it is said to be an instance of a class. The creation is described as instantiation. The instantiation uses a special type of method in the class referred to as a constructor. If you create a class but do not define a constructor, Java creates a default constructor for you.

When an object has been instantiated, the programs and data stored in the class are available for use by the object. For example, a class named `car` could be defined with an attribute called `name` and a method called `displaySpeed`. Then if an object `myCar` was instantiated, any method inside this object could access the class members as `myCar.name` and `myCar.displaySpeed()`.

Let's now look at the following program that was presented in Chapter 4.

```java
import java.util.Scanner;
/*
to illustrate input of a string and of an integer
 */
class ExampleProgram {
    public static void main(String[] args) {
int num,sum;
num = 10;
sum = 0;
System.out.println("What is your name");
Scanner nameInput = new Scanner(System.in);
String yourName = nameInput.nextLine();
System.out.println("Hello " + yourName);
System.out.println("Please enter a number between 1 and
10");
// the program will fail if an integer is not entered
Scanner numberInput = new Scanner(System.in);
int yourNumber = numberInput.nextInt();
sum = num + yourNumber;
System.out.println("Your number plus 10 = " + sum);
}}
```

In the first line:

```java
import java.util.Scanner;
```

`import` is a keyword which requests the program to use methods from the `Scanner` class. These are stored in the package `java.util`. A package is just a collection of classes. If you wished to use all of the classes in the package, you would use:

```java
import java.util.*;
```

Further down there is:

```java
System.out.println("What is your name");
```

Here, `System` is a class which is automatically available for use in a Java program (no `import` is needed).

`out` is a member and `println` is a method which has a single argument. This is not the full story here, but it is sufficient for your needs.

The following line looks confusing with the word Scanner occurring twice:

```java
Scanner nameInput = new Scanner(System.in);
```

This is an assignment statement including a declaration. `Scanner` at the beginning of the line is stating that `Scanner` is the type for the object `nameInput` which is being created.

`new` is a keyword which defines the statement as being used for instantiation.

The `Scanner` that follows this is the constructor from the `Scanner` class. In Java, the name of the class is always used to name a constructor for the class.

`System.in` is a parameter that ensures that there is keyboard input being used.

The next line:

```
String yourName = nameInput.nextLine();
```

is another assignment statement including a declaration. It supplies a value for the variable yourName. (If the variable yourName has already been declared, the String type definition is omitted.)

nextLine() is a method which is defined in the Scanner class and therefore available for use by the object nameInput.

> Glossary

abnormal data: values used for testing that the program should detect as being impossible values.

abstraction: an overview that contains the minimum necessary detail.

algorithm: a series of actions required to achieve a specific outcome.

alpha testing: the stage of testing when the development team are testing using their own systems.

argument: a value supplied when a subroutine is called.

arithmetic expression: an expression providing a numeric value to a variable.

arithmetic operator: an operator that is used in the evaluation of an arithmetic expression; for example, $+ - \times \div$.

assign: a word that applies specifically to supplying a value to a variable.

assignment statement: a line of code that uses an expression to assign a value to a variable.

beta testing: the stage of testing when the developed product has been installed on the client's systems.

Boolean logic: a form of logic that uses values that are either true or false.

Boolean operator: one of AND, OR or NOT (there are others but they will not be considered in this book).

boundary data: values used for testing that are just inside or just outside the range limits for normal data.

breakpoint: a facility used in a debugger to identify a line of the code where execution should stop.

bubble sort: an algorithm that sorts by successively comparing adjacent values.

call: the term used when a program activates the subroutine code.

call by reference: the action of providing a value to a subroutine that causes changes to the value inside the subroutine. This change is then made to the value in the calling program. Java does not use call by reference.

call by value: the action of providing a value to a subroutine that causes any changes to the value to be confined within the subroutine.

camelCase: a way of creating a variable name from a combination of at least two words. Each new word after the first word starts with a capital letter.

CASE statement: a method of providing multiple paths through the code based on the value for a single variable. It is called a switch statement in Java.

check digit: a value that is calculated from the numeric digits in a code used for identification and is then added to the code as an additional digit.

close: an action that releases the file access when the program no longer needs to use it.

command line prompt: a facility provided by an operating system that allows keyboard input and screen output associated with running a program.

comment: a description of the algorithm written within the code. The comments are intended to help explain how the code works. Comments are ignored when the code is executed.

compiler: software used to translate program code into an intermediate code.

computational thinking: a term that covers aspects such as abstraction, problem analysis, step-wise refinement, decomposition and algorithm design.

condition: the criteria that are tested which provide either TRUE or FALSE as an answer.

conditional statement: coding used to program a selection construct. This statement contains a logic proposition and the definitions of dependent actions.

constant: a named memory location used to store a value. The value can be used but not changed during program execution.

count-controlled loop: a type of iteration that will repeat a section of code a defined number of times.

data integrity: the term used to describe data that is accurate and up-to-date.

data structure: an item in a program that has a single name but can store more than one value.

data type: a specification of the kind of value that a variable will store.

debugger: specialised software used to locate errors in a program.

debugging: a general term for systematically searching for problems in programs that are not working properly and fixing them.

declaration statement: a line of code used to identify a variable name and its data type.

decomposition: a process of increasingly dividing a design into smaller components.

detailed design: the final product of a design process that contains enough detail for programming to begin.

Do while loop: the other alternative for coding a post-condition loop.

dry-run: a process where you follow the progress of a program by carrying out the calculations and recording the changing values for the variables.

element: a component of an array that stores a single value. An element can be used in a program in the same way that a variable can be used.

ELSE statement: a continuation following an IF statement which defines an alternative action.

error message: an output from the compiler caused by your code containing syntax errors.

exception handling: the inclusion of some coding that defines what should happen if the program cannot do what has been requested, such as use a specific file that cannot be found.

execute: in Computer Science, the term 'execute' means the operation of a computer program. When a computer program is in operation it is being executed. The term 'run' is also used to describe the same process. The 'program is running' or the 'program is being executed' both mean the same thing.

expression: a value or a combination of values and operators that can be evaluated.

extreme data: values used for testing that are on the limit of the range of normal values.

flowchart: a diagram used to document a detailed design for an algorithm that shows the logical flow of the actions.

For loop: a coding for implementing a count-controlled loop.

format check: a check that the data input matches a defined pattern.

function: a subroutine that returns a value to an assignment statement expression.

global variable: a variable that is declared at the beginning of a program to make it accessible throughout the whole program code.

IF statement: a statement that allows a program to follow or ignore a sequence of code depending on a Boolean condition.

imperative or procedural language: a high-level programming language used to tell a computer what it has to do and how it should do it.

index: an integer value that identifies the position of a character in a string, or an element in an array.

infinite loop: a loop that will iterate indefinitely. An infinite loop has no content that allows the loop to finish.

initialise: an action that provides a value for a variable before the variable is first used in a program.

instantiate: the action that causes memory space to be allocated for an array.

Integrated Development Environment (IDE): software that helps programmers to design, create and test program code.

iteration: a logic construct in which a sequence of actions in an algorithm is repeated several times.

Java Byte code: an intermediate code for a program that can be ported to different hardware devices.

java command: the command you use at the command line prompt to run the Java Byte code.

Java Development Kit (JDK): the part of the Java Platform that is provided for writing programs and converting them into Java Byte code.

Java Platform: the product that contains the programming language plus supporting software.

Java Runtime Environment (JRE): the part of the Java Platform that is provided for running a program using the Java Virtual Machine.

Java Standard Edition (JSE): the version of the Java Platform that is provided for developing programs to run on a PC.

Java Virtual Machine: software installed on a hardware device that runs a program by using the Java Byte code.

javac command: the command you use at the command line prompt to compile a Java program.

keyword: a word with a specific meaning defined by a programming language.

length: the number of elements in an array.

length check: a check that an input has the correct number of characters.

library routine: a subroutine provided as part of a programming language installation that can be used by any programmer.

linear search: a technique where values in an array are successively compared to a search value.

literal: a value used directly in a program.

local variable: a variable that is only recognised and usable in the block of code in which it is declared.

logic construct: a means of controlling the order in which algorithm actions are performed.

logic expression: a formal version of a logic proposition that contains a combination of Boolean values and Boolean operators that equates to an overall value of true or false.

logic proposition: an expression or statement that can be judged to be either true or false.

logical error: an error which is due to the program code having faulty logic; this allows the program to run to completion but with output that is not what should have been produced.

loop construct: a coding for including iteration in an algorithm.

loop control variable: a variable defined in a For loop header assigned an initial value. This value is then incremented at each iteration.

loop header: a line of code at the start of a For loop that defines the number of iterations.

method: the object-oriented programming name for a subroutine.

microchip: a solid-state device with installed software.

modulus: an operator that produces the remainder if one integer is divided by another integer.

nested iteration: a programming construct used with 2D arrays where a For loop contains in the loop body a further For loop.

nested selection statement: a coding structure where an IF statement is included inside another.

normal data: values used for testing that should produce sensible output.

one-dimensional (or 1D) array: an array that can be visualised as either a row of values or as a column of values. It uses a single index to identify the elements.

open: an action in a program that allows access to a file in your filestore.

overflow error: an error caused by a calculation producing a value for a variable that is too large for the storage space allocated in memory for the variable.

parameter: a variable to be used in the subroutine. It is identified in the subroutine header.

port: a word used to describe the transfer and installation of software from one system or device to a different system or device.

portability: a measure of how easy it is to transfer software from one system or device to a different system or device and make it usable.

post-condition loop: a type of iteration where the decision to repeat a section of code depends on testing a condition at the end of the loop.

precedence: a property defined for each operator to direct the order in which an expression is evaluated.

pre-condition loop: a type of iteration where the decision to repeat a section of code depends on testing a condition at the start of the loop.

presence check: a check that there has been a value entered.

procedure: a subroutine that performs one or more actions.

pseudo number: a collection of numeric digits that are never intended to be used in a numeric calculation; for example, a telephone number.

pseudocode: a way of unambiguously representing the sequence and logic of a program using both natural language and code-like statements.

range check: a check that a value falls within a defined minimum and maximum value.

regular expression: a combination of characters using a syntax specific to the programming language for defining the format for a string.

relational expression: a form of expression that contains a comparison of two values using a relational operator.

relational operator: an operator that is used to compare values. It can test whether values are equal or whether one value is greater than another value. (It is sometimes called a comparison operator.)

Repeat until loop: a coding for implementing a post-condition loop.

return: the feature, unique to a function, which provides a value to be used by a program to evaluate an expression in an assignment statement.

rogue value: a value entered by a user to stop the looping, for example, –1.

runtime error: an error that occurs when the values being used by the program lead to the program being unable to perform an action.

scope: this defines the parts of a program where a variable will be recognised and can therefore be used.

search value: a value that you are looking for in the stored array elements.

selection: a logic construct in which alternative algorithm actions are possible with the choice dependent on testing a condition.

selection construct: a part of an algorithm that allows alternative actions where the choice of which action to perform is defined by testing a condition.

sequence: a logic construct where individual actions in an algorithm follow one after another in order.

source code: a program written by a programmer using a high-level programming language.

standard method of solution: a generic technique that can be applied in many applications.

step: a value that defines how the loop control variable is incremented.

step-wise refinement: a process of increasingly adding detail to a design.

structure diagram: a hierarchical display of how a design can be broken down into smaller individual components.

Structured English: a method of documenting a design using brief English statements.

subroutine: an independent section of code that can be called from another routine while the program is running. In this way, subroutines can be used to perform common tasks within a program.

syntax: the specific words, symbols and constructs defined for use by a particular language.

syntax error: an error that occurs when the code you have written does not conform to the rules defined for the language.

text editor: software that allows you to create plain text with none of the stylistic features offered by a word processing application.

top-down design: a process where an initial abstraction is expanded to contain more detail, possibly by means of decomposition.

trace table: the table used in a dry-run to store the changing values for the variables, outputs and user prompts.

two-dimensional (or 2D) array: an array that can be visualised as a grid or a matrix.

type check: a check that the value entered is of the correct data type.

validation: the process of programming a system to automatically check that data satisfies a set of specified input criteria; for example, passwords must be longer than six characters.

variable: a named memory location used to store a value. The value of the data can be changed during program execution.

verification: a check that the data entered is what was intended to be entered.

While loop: a coding for implementing a pre-condition loop.